AF804935

Man Forest Interface
in
North East India

Man Forest Interface *in* North East India

Editors

Sumana Paul
Chandana Bhattacharjee

B.R. Publishing Corporation
Delhi-110 052

Published by:

B.R Publishing Corporation

425, Nimri Colony, Ashok Vihar, Phase-IV

Delhi-110 052

E-Mail : brpc@vsnl.com

First Published 2018

© Editors

ISBN 9789387587465

Printed at D.K. Fine Art Press (P) Ltd., Delhi

Publisher's note:

Every possible effort has been made to ensure that the information contained in this book is accurate at the time of going to press, and the publisher and author cannot accept responsibility for any errors or omissions, however caused. No responsibility for loss or damage occasioned to any person acting, or refraining from action, as a result of the material in this publication can be accepted by the editor, the publisher or the author. The Publisher is not associated with any product or vendor mentioned in the book. The contents of this work are intended to further general scientific research, understanding and discussion only. Readers should consult with a specialist where appropriate.

Every effort has been made to trace the owners of copyright material used in this book, if any. The editor and the publisher will be grateful for any omission brought to their notice for acknowledgement in the future editions of the book.

All Rights reserved under International Copyright Conventions. No part of this publication may be reproduced, stored in a retrieval system, or transmitted in any form or by any means, electronic, mechanical, photocopying, recording or otherwise without the prior written consent of the publisher and the copyright owner.

Cataloging in Publication Data--DK

Courtesy: D.K. Agencies (P) Ltd. <docinfo@dkagencies.com>

National Seminar on Man-Forest Interface: Scenario in North East India (2009 : Shillong, India)

Man forest interface in North East India / editors, Sumana Paul, Chandana Bhattacharjee.

pages cm

"The two day long National Seminar on Man-Forest Interface: Scenario in North East India was organized successfully by the academic committee, Women's College in collaboration with Office of the Principal Chief Conservator of Forests, Government of Meghalaya, Shillong on 8th and 9th July 2009"--Preliminary page.

Includes bibliographical references.

ISBN 9789387587465

1. Forestry and community--India--Meghalaya--Congresses. 2. Forestry and community--India, Northeastern--Congresses. 3. Community forestry--India--Meghalaya--Congresses. 4. Community forestry--India, Northeastern--Congresses. I. Paul, Sumana, editor. II. Bhattacharjee, Chandana, editor. III. Women's College (Shillong, India), organizer. IV. Meghalaya (India). Office of the Principal Chief Conservator of Forests, organizer. V. Title.

LCC SD387.S55N35 2018 | DDC 333.750954164 23

Foreword

I consider it as an honour shown to me to be asked to write the Foreword to the volume entitled *Man Forest Interface in North East India*. The volume comprises informative articles presented in the seminar organized by the Women's College, Shillong on 8th and 9th July, 2009.

As a seminar theme, the current status of forest and forest-based natural wealth in North East India has great importance. The manner in which forests are rapidly dwindling due primarily to human interventions invites a serious and dispassionate assessment. The day is perhaps not far off when the threshold point of nature's power to recoup and regenerate itself shall be crossed and we shall have no one to blame for such a catastrophe but ourselves. The significance of forest and need to remind ourselves of the age-old tradition of co-existence of man with the forest is more now than ever before. We all know that the traditional societies were forest-dependant societies for their survival. Yet they were careful never to over-exploit what nature gave them. Somewhere along the way, we have forsaken the adage that Man Forest relationship is one of the unique equilibria between man's need and forest's capacity to give.

The character of the relationship between the two components of the eco-system, man and forest, began changing fast with the increase of population coupled with ever-increasing demand for land and ever-expanding greed of man for forest produce, especially timber. New technologies were used, at times mindlessly, all over the world for tapping forest wealth and India too did lot lag behind in this respect. The seminar theme has rightly been chosen to discuss about forest-related situation obtaining in the north eastern part of India about 84 per cent of which was covered with forest not so long ago. We seem to be happy today if the figure is around 33 per cent of the total land area. The moot question is, do we have even that much on the ground or only in paper? Indeed a matter to make us sit up and take remedial action in a *now-or-never* mode.

The content of this volume provides valuable insights, information and recommendations. The scholars from the academia as well forest administrators with intimate, first-hand knowledge of the status of forest in the region, took active part in the deliberations. The readers will certainly be enlightened by the articles contained in this volume.

As a lay participant of the seminar, I found it to be a great learning experience. I am sure the readers of the pages that follow too will find the articles equally stimulating and useful.

Annada C. Bhagabati

Introduction

Sumana Paul
Chandana Bhattacharjee

Forest reserves are the gifts of nature. It preserves the genetic diversity of living resources. Genetic diversity is the raw-material for scientific and industrial innovation. The most important produce from forest is wood. The use of wood ranges from domestic use to that in commerce, communication, defence, etc. Despite big leaps taken in technology, man could not dispense with wood as a source of energy.

Forestry offers the rural people a livelihood. They harvest and process different forest products from timber and fuel wood to mushroom and medicines. Improving and expanding the forest resource has an important additional spin-off. Forest stabilizes the environment. On the foothills of Himalayas, deforestation happens to be the principal cause of soil erosion, which affects not only hill slopes and grazing grounds but also immensely increases the potentiality for flooding and silting up of reservoirs. The principal aim of Forest Policy, thus, must be to ensure environmental stability and maintenance of ecological balance which are vital for sustenance of all life forms. However, despite the fact that forest management in India had a glorious past (which dates back to several centuries Before Christ), in recent times Indian forestry is facing a number of challenges.

Forests supply timber, firewood, fruits and among other raw materials for industries. Through transpiration forests supply moisture to the atmosphere and contribute to the increased rainfall. Forests are powerful checks against soil erosion. The fertility of the soil is increased by the addition of decomposed forest products. Forests also control floods and the velocity of the air current. By drawing carbon-di-oxide forests purifies atmospheres. Forests shelter animals are only a few of the merits of forests to mention.

Since 1960, with the flourishing of wood and ply-wood industries in India, forest cover of the country has been fast reducing. There are several continuing examples of many small communities in our country taking self-propelled initiatives outside the formal structures of law and governance to protect their forests. In their endeavor to safeguard their environment by protecting forests, they have often run into direct conflict with powerful political and economic structures. The same forests have different interests for others. This results in the destruction of forests at an alarming rate. Deforestation happens to be one of the major contributing factors to global warming. Many environmental historians are of the view that the large scale deforestation in India is rooted in the commercially oriented forest use. In more recent times rapid industrialization, urbanization and growing consumerism have led to wide-scale destruction of forests.

India's north eastern region is bestowed with immense natural heritage. It is one of the richest bio-diversity hotspots in the world with a wide variety of flora and fauna, some of which are solely endemic to this region. With forests ranging from the Alpine belt of the eastern Himalayas to the lush green tropical jungles and the swampy grasslands, this is a nature lover's paradise. Unfortunately at present the forest cover of this region is only 64 per cent of the total geographical area which was as high as 84 per cent only two decades earlier. Some of the species of animals, birds, and plants, found uniquely in this region, are highly endangered.

The conflict arises when man imposes upon the natural functions of the forest. This has resulted in an unending, complex debate. The complexity aggravates due to intricate relationships which exist between the functioning of the man and the forest, the multiplicity of functions that the forest perform (related to qualities) and the interference of man with their ever increasing needs and attributes (related to quantities). While forest areas in India, as anywhere else in the developing world, have considerably reduced in the last few decades for various reasons, the demand for forest has always been increasing.

The areas of the forest is reduced for developmental reasons, by encroachment using it as grazing ground for domestic animals, tree felling, etc. all of these disrupt the natural functions of the forests. The consequences are adverse in every sphere as evident. The National Forest Policy, 1952, Forest (conservation) Act, 1980, aims at maintaining forest land and ensures the sustainable development of the forest respectively. Despite these policies and Acts, between the periods 1951-1980 shows the staggering figure (2623 thousand hectares) area diverted for agriculture alone in our country.

The loss of forest is causing hazards all over the world, allover India, and certainly north eastern part of India is no heaven in this context. B. K. Talukdar, Secretary General of AARANYAK that works for biodiversity conservation, said that (June 23, 2008, New Delhi) their latest findings on the forest cover of the north east India, clearly shows that 64 per cent of the total geographical area of this region (255,000 sq. km.) is currently under forest cover. Two decades earlier, it was 84 per cent. This clearly sends the message to its entire people to be concerned and act promptly to retain the existing forest from any further damage, work together to enhance its resources.

Participatory Forest Management, though a fairly recent concept, is gaining appreciation in India. Implementing the development-based programmes involving the local people wherever initiated by the enthused forest personnel, have yielded exciting results. This idea of peoples' participation is based on the realization that no such programme can succeed without the active support and participation on the part of the people.

The contribution to forestry can fully be realized through the effective management of the forest. The purpose of the management certainly should include rural development aspect. The rural populace must not be deprived from the benefits that flow from the forest wealth. Minor Forest Produces provide sustenance to the tribal as well as other communities residing in and around the forests in north east India as in any other part of the world. Such produce, therefore, calls for protection, improvement, and their production enhanced.

Indiscriminate felling of trees and large-scale soil erosion due to heavy precipitation leading to severe water scarcity at Cherrapunji has drawn world-wide attention. The land, once with green forests, now known as 'wet desert' because of the fact that the bare hillock have already washed away the limestone rich soil at the rate of 300 tons per year. Alarmed with the barrenness of world's wettest place, the local Government has proposed a project to're-green' Cherrapunji.

II

The two day long national seminar on Man-Forest Interface: Scenario in North East India was organized successfully by the academic committee, Women's College in collaboration with office of the Principal Chief Conservator of Forests, Government of Meghalaya, Shillong on 8th and 9th July 2009. This seminar was a part of the Silver Jubilee Celebration of this esteemed institution. The volume is a collection of the academic papers received and presented in the said national seminar.

The main objective of the seminar was to uphold the prevailing scenario of man-forest interface in this part of the country and to create mass awareness towards maintenance of environmental stability and restoration of ecological balance without making serious compromises on the developmental needs of the region. The deliberations of the two day seminar have realized the various objectives of the seminar to a large extent. The seminar also focused on the diverse uses of the various forest products. With this end in view, the seminar was made a livelihood oriented exercise besides being merely academic discourse. Financial assistance for holding the seminar was mainly sought from the Indian Council of Social Science Research, North Eastern Regional Centre (ICSSR-NERC) and North Eastern Council (NEC) Shillong.

The inaugural session of the Seminar commenced at 9:30 AM on July 8, 2009 at the college auditorium. The inaugural session was chaired by Dr. B. B. Dutta, President, Governing Body of the Women's College. The chief guest for the occasion was His Excellency Sri R. S. Mooshahary, Governor of Meghalaya. His Excellency, in his address, focused on the value of forest and ecological balance and the need for plantation of more trees in order to save the Earth. He also spoke on *Green Gold* that is Bamboo as an important resource of the region. The guest of honour for the occasion was Professor Annada C. Bhagabati, Former Vice Chancellor of Rajiv Gandhi University, Arunachal Pradesh, whose speech revolved round the reverence paid to nature by the communities of the region. The welcome address was delivered by Sri Debasish Chowdhury, Principal, Women's College. The Key Note Address was delivered by Sri V. K. Nautiyal, Principal Chief Conservator of Forests, Department of Forests, Government of Meghalaya. He stressed on the theme of the conference and discussed the important role played by forest in the lives of Indians. He also stressed on the various forest policies formulated by the Government of India. He also focused on the unique forest resources and biological diversity of the region and the challenges the region is facing. In the Key Note Address delivered by V. K. Nautiyal, some important dimensions have emerged out. While looking at the entire scenario of the north east region of the country he has historically linked together the various facets of our country by way of portraying the man-forest dependence and its relation to the society within the broad framework of developing parameters that the country has been experiencing over the long passage of time. At the same time he has also hinted at the current trends of man-forest interdependence *vis-a-vis* the factors that have been degrading the rich natural heritage of the nation. The Government Forest Policies enacted since the colonial times have acted as guiding principles for the people at large. He concludes with ten touchstones that can be viewed as a flagship programme to understand the man-forest interface of the country in general and of this region in particular.

The academic sessions began after the inaugural session. Keeping the broad theme of the seminar as a referral point, several sub-themes were worked out. Papers under these sub-themes were presented and discussed, which focused on creative integration of traditional knowledge based technological development that positively influences improvement of livelihood of shifting cultivation practicing communities of north eastern India; potentials of north east India in terms of its rich bio-diversity *vis-a-vis* the persuasive measures of sustainable development; the practice of shifting cultivation attributing to the loss of valuable natural resources of the state of Meghalaya; tradition based dying method; development of industries for the production of textiles using natural dyes, which could provide a solution to the problem of unemployment as it enjoys a profitable market in India and abroad; non-integration of Biotechnology research with the problems and constraints confronting environment, forest resources, sustainable development and the low-income farmers in the agricultural sector; problems within sustainable agriculture and rural development that have historically been linked to conventional technologies; regional potential in developing fast growing tree species, varieties for energy plantations and stress tolerant plants and cross-condition adaptable crops to benefit the population; Joint Forest Management, including farmers' groups in decision-making regarding the definition of sustainable agriculture and rural development objectives and defining proper forest policies flawed government policies responsible for destruction of our forests and the environment; alternative framework for a socio-environmental movement; ideas of a movement which would be essentially social in character; Land and Forests rights of tribals in West Khasi Hills of Meghalaya; crisis in man-forest relation; serious concern with the opinion that educational intervention through environmental education programmes in formal and non-formal settings can be an important remedy to minimize the crisis in man-forest relation; conserving and managing the forests of the region for our sustainable use requiring a massive drive to make the people aware about the seriousness of the situation and their duties and responsibilities to avert the impending disaster, which can be done through education, media publicity, public meetings, organizing workshops, seminars, etc. where students, public leaders and the women folk are felt to be the priority target groups in such campaign, who will be able to create a lot of impact on the society; human-animal conflict in north east India in general and in Assam in particular with an animal corridor with habitat and provision of adequate space with food, allowing the pray base thrive and multiply, conflict will stop or minimize to the great extent, else, animals will continue to find ways of getting food from humans, raiding of paddy fields, ransacking of human habitats, lifting of cattles; climate features and biodiversity concern; vicious circle of poverty, sustainable development of the economy should be the approach which takes into consideration both economic and ecological attributes while going forward with the developmental plans; human intervention: effects on forest and wildlife in Meghalaya; and significance of increasing cultivation of Bamboo grass which can be properly utilized for sources of various types of livelihood.

The seminar brought forth the following major recommendations:

- There is an urgent need of more and more tree plantation in the possible areas like roadsides and barren lands.
- More and more plantation of Macaranga tree can be a solution to the wide scale deforestation
- There is a need of passing on traditional knowledge to future generations for the conservation of natural resources in the region.
- A creative integration of traditional knowledge based technological development will positively influence improvement of livelihood of shifting cultivation practicing communities of north-eastern India.
- Innovative method of cultivation can be made more widespread to make it more sustainable and environment friendly.
- Cultivation of Bamboo grass should be encouraged on account of its diversified application.

- There should be proper initiative on the part of the govt. and individuals to promote local markets for indigenous textile.
- Man- animal conflict may be avoided by showing little sense of sympathy, empathy and understanding towards them.
- Educational intervention through environmental education pregrammes in formal and non-formal settings can be an important remedy to minimize the crisis in man-forest relation.
- To prevent the crisis of man-forest relation Peoples participation is essential.

III

We would like to record our special thanks to His Excellency Governor of Meghalaya, Sri R. S. Mooshahary, for kindly gracing the occasion as the Chief Guest and inaugurating the national seminar. We are indeed grateful to Professor Annada C. Bhagabati, former Vice Chancellor, Rajiv Gandhi University being present on the occasion as the Guest of Honour. We also express our sincere thanks to Sri V. K. Nautiyal, former Principal Chief Conservator of Forests, Department of Forests, Government of Meghalaya for delivering the Key Note Address. While the contributions were being arranged for bringing into a final shape in publishable form we got academic impetus and assistance from Dr. Gautam Kumar Bera, a senior social-cultural anthropologist, to whom we are also indebted to. We are thankful to North Eastern Council and Indian Council of Social Science Research (NERC), Shillong for their financial support to make the national event a grand success. Last but not the least we acknowledge our sincere thanks to Doordarshan, other electronic and print media of Shiilong for their coverage.

Contents

1

Man Forest Interface: Scenario in North East India

V. K. Nautiyal

Forests have played an important role in the life of Indians right from the ancient times. The Dravidian Civilization in the southern part of India apparently lived in consonance with thick extensive forests and its wildlife. The Aryans were worshippers of nature and they had their adobe and educational centers in sylvan surroundings and inspiring landscapes. Emperor Ashoka, the far sighted monarch, had envisaged that useful trees be planted along the roads and on camping grounds. From the eighteenth century onwards the population of the country increased rapidly and as a result of this there was extensive clearing of forests for agriculture and for other human needs.During the Colonial times, forests were harvested on a large scale for providing wood for building ships for British navy and for railway sleepers, that helped in extending the railway lines. The scientific management of forests started with the establishment of Imperial Forest Department in 1864 and Dr. Dietrich Brandis, a German botanist was appointed as the first Inspector General of Forests. In 1865, the Indian Forest Act was drafted which was amended several times to assume its final form in 1927.In 1894 the Government declared its first forest policy which envisaged that forests situated in hill slopes be maintained as protection forests to protect the cultivated plains that lie below them. The forests having valuable timber were to be managed on commercial lines as a source of revenue to the State. As per the policy, the demand for agricultural land was to be met from the forest. The forests that yielded inferior timber for wood or for grazing were prescribed to be managed mainly in the interest of local population. However, the state ownership of forest lands was enforced through rights and management of degraded forests at the periphery by the local community was recognized. In 1952, The National Forest Policy was revised which provided against indiscriminating extension of agriculture by excision of forests. The national forests were to be managed on the principle of higher sustainable yield to meet the requirements of defense needs and industries. It emphasized the need for ensuring supplies of small timber and agriculture implements and fire woods etc. for the public. The National Forest Policy was again revised in 1988, the salient point of which was to maintain basic needs of the people and strengthening of tribal forest linkages that were to be given due importance. Conservation was the main priority and earning of revenue took a back seat.

Despite the rapidly growing population and pressure on forest lands for meeting the demands of people and industries, the forests in India have a unique place in the overall scenario in the world. The total forests cover of the country as per the State of Forests Report of 2005 is 6, 77,088 Square kilometers. Which constitutes 20.60 per cent of the land area. India is one of the 12 mega biodiversity countries that collectively contain 60per cent of the entire biodiversity of the world. Sixteen forest types are available in the country which is the home to 75,000 species of animals and 45,000 species of plants.

The north east region of India is in some ways akin to the rest of the country while in many other ways it is quite different. Within the north east region itself the states are very different from one another. To complicate the situation further, even within the states itself there are great ethnic, cultural and linguistic variation. The region originally included the seven states, east of West Bengal but of late the State of Sikkim is also included in the North east.

As per an old report, north east has 7.76per cent of the countries' geographical area but 26per cent of the forest cover and 31per cent of the growing stock. With the inclusion of Sikkim, it now encompasses 2, 62,179 Square kilometers area with a population of 39 million as per the census of 2001 and more than 70per cent area is hilly. The population density varies widely. Assam and Tripura are most densely populated states with more than 300 people per Square kilometers, which is around the average for the country while Arunachal Pradesh has only about 13 people per Square kilometers. Hence the north eastern region is characterized by the great diversity of people.

However, of equal importance is the biological diversity of the region. The north east is at the confluence of Indo-Malayan and Pale arctic biogeographical realms. The WWF has identified the entire Eastern Himalayas as a priority global 2000 Eco- region and Conservation International has subsumed its Eastern Himalaya "hot spot" into a wider Indo Burma hot spot which now includes all the eight states of the north east. The region is one of the endemic bird areas defined by Birdlife International, harbours and IUCN centre of endemism. It is also an important sub center for the origin of cultivated crops which means that many wild botanical species of the region are progenitors of the food plants. The north eastern region supports 836 of the 1200 bird species known from the Indian sub-continent reflecting the richness of diversity of habitats. Assam hosts the entire known population of pygmy hogs and 75per cent of the world population of the Indian Rhinoceros and wild water buffalo. Then there are large population of wild elephants more known to common people for the problems they create. In addition there are variety of cat species and macaque species. The region is specially rich in insect life as large scale use of pesticides is not practiced.

The situation of forests vis-a-vis people in the north east has thrown up several challenges. The first and foremost is how to conserve the biodiversity in the complex socio-political and ethnic scenario. One of the main conflict situations has arisen due to increasing demand for timber in the rest of the country for meeting the requirement arising out of rapid development the nation has embarked upon. With vastly improved infrastructure in the north east, especially the road network, many areas which were earlier out of reach for commercial exploitation have now become extremely lucrative to log. At the same time the population of North east has been increasing steadily. From a total population of 1,73,325 in 1901 the population increased to 31,71,168 in 2001.This has not only put a huge pressure on forest land for diversion from agriculture but has also increased the local demand for fuel, fodder and smaller timber etc. Another important factor is shifting cultivation or swiddening. From 1987-1997, nearly 13per cent of all forests were under jhum cultivation. In the region, there is no clear cut demarcation of forest and agricultural land. Some view jhumming as agriculture on forest land while others view it as forests on agricultural land. None the less there is an overwhelming view of conservationists that jhuming is destructive land use and is not of good value for labour. Environmentalists and planners are also seized with the problem how to conserve the rich bio resources of the region. There is also the issue of the enormous hydro-electric potential of the north east which is estimated to be 60,000 mega watts of which only 2004 mega watts is developed or under construction. Damage to forest cover would affect the viability of many of these projects. Here it would be relevant to mention that 1/3rd of India's runoff flows from the north east through the Brahmaputra and Barak rivers. Large tracts of land especially in the hilly part of the region have also been diverted to mining for coal and limestone. Lush greenery of several parts of the area has been devastated by cutting of even small size trees and brushwood to make charcoal to feed ferro alloy industry.

These all and related issues have to be addressed against the backdrop of distributed socio-political conditions of large part of the North east which is aggravated by the existence of 4,800 Kilometers of porus international borders with four countries. The efforts of the Government agencies to manage the forest in the north east have had mixed results so far. While the areas which are under the direct control of the Government in the form of Reserved forests, Protected forests, National Parks and Wild Life Sanctuaries, etc have good conservation status, the areas under the control of various traditional bodies and constitutional councils have rather poor records of management. One big problem has also been that a universally accepted definition of forest has not been formulated so far and the efforts of the Supreme Court to enforce the forest conservation act vis-a-vis an adhoc definition of forests has limited success. There have been several efforts in devising community based management system. The most notable amongst these have been the scheme of Joint Forest Management under the umbrella of Forest Development Agencies involving village communities in forestry operations and protection of forests. This programme is a follow up to the new Forest Policy of 1988. The Joint Forest Management approach provides communities with a share of timber and non timber forest products from government forest in return for forest protection. These policies and programmes have more relevance to the communities living in the periphery of government owned forests in plain areas but would have limited effect in hills where most of the forest areas are outside the ownership of the Government. In such areas the Forest Department has involved the communities in creating forest cover on community land with a promise of sharing of usufructs on maturity thus ensuring public participation. There have been other non Government community based forest conservation efforts but these have been isolated examples and they have not had any wide scale effect. One disturbing trend in the region has been the dilution of traditional institutional authority resulting in weakening of indigenous forest conservation mechanisms leading to forest fragmentation, degradation and loss. At the same time there have been incidents of privatization of once community held forests and water sheds, leading to their deforestation and conversion to other land use. This has amongst other things also created socio-economic disparities in otherwise egalitarian forest dwelling societies of the north east region'.

Thus the objective of any study or plan or programme should develop on critical issues some of which I am to flag as follows:

1. Management of commercial forests especially on community land in the north east region with community participation making the communities as stake holders.
2. Increasing the productivity of the forests especially in the matter of providing fuel wood, fodder and other NTFP.
3. Conservation of bio-diversity of the region and bringing as much of the critical areas as possible under the direct control of the Government or Government agencies.
4. Developing incentive based programmes for conservation of areas having high bio-diversity values like sacred groves by involving communities.
5. Providing gainful forestry based employment to people addressing the gender based issues at the same time.
6. Reducing the areas under shifting cultivation by increasing settled cultivation of food crops in order to provide food security.
7. Encouraging cultivation of eco-compatible cash crops in a manner that do not affect wildlife habitats and do not result in man- animal conflict.
8. Establishing linkages with Horticulture Department to promote cultivation of fruit trees and plants and floriculture.
9. Encouraging cultivation of quick yielding forestry crops like indigenous bamboos, medicinal plants, etc.
10. Cautious development of eco tourism. Half of all tourists in the world are nature tourists and contributes $514 billion annually.

With these words I wish to draw kind attention of all the participants in this seminar cum workshop on the various issues that need to look at. For the north east a symbolic relationship of people with forests is a win win situation for all. I hope this seminar would be able to bring to the fore some of the problem of modern times.

2

Traditional Knowledge Based Natural Resource Management Systems of North Eastern India

B. K. Tiwari and M. B. Lynser

Most aboriginal tribes of the world have been living in proximity to the natural surrounding and their inherent dependence on these resources for subsistence as well as cultural and health-related goods and services has enabled these communities to gain an understanding about the resources around them. Based on this understanding they developed and practiced various management techniques which helped in the conservation and judicious utilization of these resources. Popularly known 'traditional knowledge', this cumulative body of knowledge or management practices, sometimes in the form of beliefs, is handed down through generations. The traditional knowledge plays an important role in the management of complex adaptive social–ecological systems and contributes effectively in providing benefit and enhancing livelihood of the communities. It links the society with the resources and helps in the maintenance of the balance between production and extraction and thus ensures perpetuation and continued existence of natural ecosystems (Gadgil et al, 1993). There is a great variability in the management practices of traditional societies as these practices have evolved in a definite set of bio-physical and cultural milieu. Some of these practices have helped communities to survive in extreme inhospitable environmental conditions of high altitudes like the Central Himalayas ((Farooquee et al, 2004). Traditional management practices of natural resources attracts greater attention because if and when it is combined together with scientific knowledge it can potentially play a far more positive role in sustaining biodiversity and nature's services and also at the same time increases productivity of the systems involved (Becker and Ghimire, 2003; Tiwari, 2005).

North-eastern India is the home to immense and colorful cultural (with over two hundred ethno-linguistic communities), as well as natural (a biodiversity 'hotspot') diversity. The forests where most of these biodiversity is found is bestowed with abundant natural resources, about which the traditional societies of the region through use and management, have gathered a rich wealth of knowledge associated with conservation and sustenance of these resources. Such knowledge along with the strong democratic traditional institutions has ensured the continued existence of natural forests, agro-biodiversity, unpolluted streams and springs, which is in sharp contrast with the rest of the country where a majority of the forests now exist only as government notified/protected areas. The present paper documents and analyses the traditional knowledge based natural resource management systems for forest and land resources by the people of the region.

The Region and its People

The North-eastern region of India comprises 8 states, namely Arunachal Pradesh, Assam, Manipur, Meghalaya, Mizoram, Nagaland, Sikkim and Tripura. The total geographical area of the region is 2, 62,159 square kilometers accounting to 8 per cent of the country's geographical area. About two third of the land area is upland whereas only one third is lowland. The region has a population of about 40 million (Anon, 2001) and is inhabited by a diverse array of communities. Over 240 distinct ethno-linguistic communities are distributed through the mountains, plateaus, upland valleys and river plains making the region one of the cultural hot spots on the planet.

The region is endowed with immense biodiversity in vast array of forest, aquatic and agroecosystems. Forest cover in most states of north east region is above 75 per cent except for the state of Assam (35per cent) and Sikkim (47per cent) (FSI, 2009). The region harbors about 50 per cent of India's floral and faunal wealth. Fifty one forest types are found in the region broadly classified into six major forest types *viz.*, tropical moist deciduous forests, tropical semi evergreen forests, tropical wet evergreen forests, subtropical forests, temperate forests and alpine forests. The richness and variability in vegetation ranging from temperate to tropical is due to diverse topography and variation in rainfall, soil and temperature. The region has been recognized as a biodiversity hotspot. It forms a part of the two global biodiversity hotspots *viz.,* the Himalayas and the Indo-Burma out of the 34 hotspots recognized globally. The economy of most of the states in the region is agriculture and forest supplemented by limited horticulture, animal husbandry and crafts production.

Typology and Institutional Arrangement for Management of Forests in Meghalaya

The rich forest resources of the region are subjected to tremendous pressure due to various anthropogenic activities such as mining, shifting cultivation, timber trade, etc. However, in many areas large tract of dense and well protected patches of community owned primary forests can still be seen. This can be attributed to the traditional forest management practices followed by the tribal communities since time immemorial. Under customary laws these forests are classified in different types depending on the intended use whether for protection of water sources and biological resources (safety forests) or for extraction of timber and fuel wood for day to day needs (supply forests). In Meghalaya, these forests are locally known as *Law Kyntang* (Sacred Forest), *Law Shnong* (village forest open for use by the residents of the village), *Law Adong* (village restricted forest), *Law Raid* (Forests belonging to a group of villages open for use by the residents of a cluster of villages), *Law Ri-Sumar* (community forest), *Law Lum Jingtep* (cemetery forest) and *Law Kur* (clan forest). Such traditionally managed forests provide a variety of forest product (Tiwari and Kumar, 2008) as well as services *viz.*, protection of water source, protection of watersheds, conservation of biodiversity and in particular the medicinal plants.

Strong traditional institutions are prevalent in the region. The tribal people of the region manage these forests through these traditional institutions. These are the local administrative units which look after the well being of the communities as well as management of natural resources of the area under their jurisdiction. Traditional institutions manage and control their territory according to local customs and traditions. Their main task is to formulate rules and laws for smooth administration and for managing the common property resources (e.g., forest and land resources) and oversee the effective implementation of these norms. For management of the forests, very clear terms and conditions are laid down regarding the access and use of resources. Normally, violators of the regulations have to pay fixed penalty in terms of kind or cash, but sometimes the traditional institutions have flexibility and punish the offender depending on the nature and gravity of the offence committed. Subject to the conditions laid down by these institutions, the communities enjoy rights to access and use the community forests and other common property resources. They can collect fuel wood, fell trees for construction of houses, collect wild fruits and vegetables, ornamental orchids and medicinal herbs and can do quarrying of sand and stones from permitted sites.

Traditional Health Care System of Meghalaya

Tradition of health care based on folk medicines is widespread and popular in Meghalaya. The tribal communities are very knowledgeable about the wild medicinal plants and depend on the herbal product for treatment of most of their common ailments and diseases. Medicinal plants mostly collected from the natural forests are the most vital resource for the traditional health care systems. A rich knowledge on the medicinal use of plants by the different tribes was also seen from the survey conducted in the four Agro-Ecological regions of the state. A total number of 58 wild and domesticated medicinal plant species were used by the people living in the Western Region, 103 in the Central and Upland region, 52 by people in the Northern Undulating Hills and 85 species in the South Precipitous region of Meghalaya. The levels of collection of Wild Edible have been recorded to be very high in all Agro-Ecological Regions of the state (Tynsong et al., 2006).

Meghalaya has a large number of herbal practitioners. Almost every village has one or more such practitioner. As many as 70 per cent of medicinal plants used for the traditional health care systems in the state come from natural forest (village restricted forests and village forests), 10 per cent from forest gardens and 20 per cent from home gardens. Folk knowledge related to medicine and health care exists in almost all the rural communities of the state and also support the livelihood for many people. Majority of the people of the state use some form of folk medicine which includes housewives and elders, traditional birth attendants, herbal healer and bone setters. The use of such knowledge and herbal ingredients in the treatment of common ailments and in some cases even major diseases or chronic ailments cut across social and economic strata.

Safety and Supply Reserves of Meghalaya, Mizoram and Nagaland

In the state of Meghalaya, Mizoram and Nagaland the traditionally managed forest can be broadly grouped as safety and supply reserves. In the case of supply reserve, forests are a common property of the people and are usually set aside to meet bonafide needs of the communities. The communities can collect both timber and non-timber forest products (NTFPs) from these forests. In most places collection of timber and NTFPs like fuel wood is restricted to personal use only and not for commercial purposes. This is usually to avoid overexploitation of natural resources. Although, NTFPs like mushrooms, wild vegetables, fruits can be collected for sale in local markets. Traditional institution like the village durbars in the case of Meghalaya has the responsibility to ensure sustainability of these forests`(*Law Shnong*). To do this, at times a certain period is fixed; say 5 or 10 years when a forest will remain open for extraction of forest products. After that period is over, extraction from this forest is closed to allow regeneration and another forest is declared open for the extraction. In this way, the regeneration of the forest is ensured and the resource is conserved without affecting the availability of forest produce for the people.

In Meghalaya, the village restricted forests (*Law Adong)* serve as safety reserve. These forests have got higher degree of protection and access to forest resources is not open to all. These forests are especially reserved for the poorer section of people in the village and for some occasional needs by the village as a whole. Extraction of timber and fuel wood is usually restricted from such forests but for NTFPs like mushrooms, edibles fruits and vegetables can be extracted without affecting the health of the forest. Extraction from such forests is allowed only when acute needs arise like for construction of houses for the poor and needy people, for making coffin in case anybody from the village dies, for construction of community hall, school etc or in case of natural calamities. The decision for extraction of timber from such forests is taken in the 'village durbar'. Mature trees are usually harvested for timber using selective felling method. In certain cases it has been noted that mature trees are extracted for raising funds for the village or in a few cases it was also noted that the revenue generated from the proceeds is shared equally among the villagers. In all such cases the integrity of the forest is maintained and under no circumstances the forest is converted into other land use.

Many villages of Mizoram have such category of forests (Malhotra, 1990). They are managed by the village councils under over all control of Local Administration Department of Government of Mizoram. According to latest statistics 645 squere kilometers is under safety reserve and 1164 squere kilometers is under supply reserve. Likewise, most villages of Nagaland also have forests allocated for use and extraction (supply forests) and the others set aside for posterity as village protected forests equivalent to the safety forests.

Forest Gardens of War Area of Meghalaya

Tribal people living in southern Meghalaya where the weather and topography is not conducive for agriculture due to extreme rainfall (>6000mm) and steep slopes (40-60°) have developed unique production system of the forest gardens where economically useful trees are promoted within the natural forests (Tiwari, 2005). These agroforests are rich in biodiversity and provide high level of productive benefits (Tynsong, 2009). The forest gardens are probably the best land use system possible in such fragile ecosystem where once the land is made barren regeneration of forests may not be possible as is seen in Cherrapunjee plateau.

Betel Leaf Cultivation in Meghalaya

In Meghalaya, the betel leaf is grown on slopes inside forest or under the shade of trees. The growers prune the tree canopy at right time for making the openings for light. In some places the growers make bamboo channels for irrigation of the betel vines planted at the base of the trees also known as bamboo drip irrigation. The betel leaf growers observe strict hygiene of self, the implements and tools used for tendering of the plant and harvesting of leaf. Thus they prevent infection and maintain the plantation disease free without using any pesticide. Within few years the trees sprout and develop a thick canopy and provide support and desired shade to the betel vine. These plantations provide most forest ecosystem goods and services while also providing economic return to the planters. The cultivation of betel leaf inside forest without clear felling the forest is probably the most sustainable agroforestry practice evolved by the people in fragile slopes of southern Meghalaya that receives very high rainfall.

Bamboo groves of the Jamatia of Tripura

The Jamatia tribe of Tripura are a forest dwelling community and have during the course evolved sound community based bamboo management practice. The most dominant bamboo species nurtured and cultivated by Jamatias is Muli bamboo (*Melocanna baccifera*). Other species of bamboo include *Bambusa balcooa* (Barak), *Bambusa polymorpha* (Bari) and *Bambusa tulda* (Mritinga). People mostly women maintain the bamboo for house construction, repair, or making of fencing, firewood and for cash income. The bamboo is collected during the months of November and December. The number of bamboo to be extracted per family or household each year is based either on minimum average requirement of the majority of households or on consensus during a meeting of all households managing the bamboo forest. The villagers through their knowledge and experience decide the number to be extracted each year ensuring sustainable yield.

Sacred Forests of Meghalaya

Many fully developed, virgin primary forest patches are seen in and around human settlements at various places in Meghalaya. Protecting of such patches of forest on religious and cultural ground is an age old practice in Meghalaya. These patches of forest are known as Sacred Forests or Sacred Groves (SGs) and are called by different names in the state like *Law Kyntang, Law Lyngdoh and Law Niam* in Khasi, *Khloo Blai, Khloo Blai Lyngdoh* in Jaintia and *Kanggimin Bol-Waarangni Biap and Asang Khosi* in the Garo Hills. The indigenous people of Meghalaya believe that their deities live inside these SGs and would be offended if any damage is caused to the plants and animals in these SGs. Hence nobody collects anything from the forest, not even fallen branches of trees. Trees like *Castanopsis tribuloides* var. *ferox* (Fagaceae) are not allowed to be cut by the local people. The responsibilities for the maintenance of SGs lies with the village council or the clan council to which it may belong. According to the State Forest Department, SGs cover an approximate area of 1000 square kilometers in the State. A study by Tiwari *et al.*1999 have documented 16 SGs in Garo Hills, 48 in Khasi Hills and 15 in Jaintia Hills. These groves are known as a treasured house for biodiversity and rendered innumerable ecosystem services. They also serve as micro-level biodiversity hotspots and gene bank of the ecosystem in a degraded environment.

Traditional Knowledge and Innovation in Shifting Cultivation

The most prevalent and age old land use practice in the sloping lands of north eastern hill region is Shifting cultivation. This type of cultivation is mainly practiced in common or community land and it is closely linked with the tribal culture in the region (Syiemlieh, 2007). Presently more than 0.45 million tribal families in the region are engaged in this type cultivation. Way back in 1961, the Dhebar Commission, Government of India estimated that shifting cultivation in north east India was practiced in an area covering 9,209 square kilometers. However, recent studies have shown that the area under shifting cultivation has increased to 19,820 square kilometers which accounts for 87 per cent of the total shifting cultivation area in the country (Satpathy et al, 2003).

Creative integration of traditional knowledge based technological knowledge in shifting cultivation practicing communities of north-eastern India leads to livelihood improvement, conserve biodiversity as well as land resources. Shifting cultivation on community land can usually be seen in small patches only as farmers are not allowed to clear more land than what they can work. If a farmer cannot use the land allotted for shifting cultivation by the community within 3 years, the land reverts to the community again. This helps in preventing misuse of land and forest resources by individuals.

Shifting cultivation is done mainly for meeting subsistence requirement rather than market demand. A diversity of crop altogether can be grown within a small area taking into account the cultivators family needs, likings and constraints. For example in Umden, Ri Bhoi district crops like rice, chilly, pumpkin, ginger, black sesame, yam, sweet potato can be grown in a single plot. From health point of view, crop products from shifting cultivation is free from chemicals hazards, as application of chemical fertilizers are usually avoided by people for fear of deterioration of soil health over a period of time.

Konyak Type of Shifting Cultivation, Nagaland

Inhabited by Konyak people, Ngangching village is located 42 km south of Mon town, the headquarters of Mon district. About 30 per cent per cent of land resources in the village are under primary forests and the rest 70 per cent are under shifting cultivation and other land uses. Traditional shifting cultivation practiced by the Konyak people is done in a balanced way so as to serve the twin objective of livelihood and conservation of resources.

The most unique and special feature of the *Konyak* type of shifting cultivation is the fact that they maintain trees as a 'crop', nurturing them for the next *jhum*. Trees from previous jhum cycle(s) like *Trema orientalis*, *Sapium baccatum*, *Grewia* spp., *Quercus* spp , *Alnus nepalensis* and *Schima wallichi* poles are lopped at a about 3 metre height. In current jhum fields, growth of *Macaranga denticulata* is promoted through selective weeding, spacing, and pruning. Konyaks revere *Macaranga denticulata* because of its varied uses. As the first slash for jhum is to be undertaken, they plant this tree and declare "go away the bad, come all the good". This species is also used in rituals to dispel evil spirit that is causing sickness to a person. They believe that the patient would recover immediately or die that day when a ritual is performed with this plant.

Trees of *Cedrela serrata, Duobanga grandiflora, Melia* spp., *Spondias axilaris*t are not lopped and allowed to grow to its full height. These unlopped trees are preserved for house construction as main posts and also timber extraction. Planting of perennial cash crops viz., banana and large cardamom underneath *Macaranga* trees is commonly observed. *Lavistona jenkisiana* (Toko) gardens are an important element of the landscape as the leaves of this plant are used for thatching. Steep escarpments and ravines are left uncultivated to serve as *in situ* conservation of biodiversity and source of wild food.

Discussion

Traditional knowledge helps in conservation of natural resources. It provides social, food and environmental security as it takes care of soil, water and biodiversity conservation, food security and health care of the people. It can also be seen that traditional management systems play an important role towards ensuring sustainable livelihood of the tribal communities both in short and long run. The systems are build upon active participation of the people and equity and social justice is key objective of the management. This strengthens social bondage/cohesion and reduces conflicts between different groups of people. The management system need little external inputs, is flexible and evolves with time for which an in built mechanism is in place. Communities are constantly innovating and improving their production systems. Natural resources and traditional knowledge are interrelated and interdependent for perpetual existence so we should put efforts to judiciously blend traditional knowledge with modern principles of conservation and management of natural resources as this will further strengthen conservation issues.

Traditional knowledge based natural resource management is based on the wisdom held and shared by the community, and passed down from generation to generation. For a long time, traditional management practices were ignored by modern science. It was said to be primitive, superstitious, or unscientific. However, these knowledge systems or traditional practices of different communities are so valuable that they are the source of solution of many present-day problems. They are often more appropriate for the local environment and conditions than solutions or methods suggested by outside experts. With the world changing rapidly, traditional management systems can be easily lost, so it is important to make efforts to preserve it.

References

Anon, 2001. Basic Statistics of NER 2006. (online: http://databank.nedfi.com/content/general-information) retrieved on 2.12.2009.

Barik, S.K., Tiwari, B. K and 2006. *Sacred Groves of. Meghalaya –A Scientific and Conservation Perspective*, Regional Centre, NAEB, NEHU, Shillong.

Becker, C.D., Ghimire. K. 2003. Synergy between traditional ecological knowledge and conservation science supports forest preservation in Ecuador. *Conservation Ecology*, 8(1):1 (online:URL: http://www.consecol.org/vol8/iss1/art1/) retrieved on 25.5.2007.

Farooquee, N. A., Majila, B.S and Kala, C.P. 2004. Indigenous Knowledge Systems and Sustainable Management of Natural Resources in a High Altitude Society in Kumaun Himalaya, India. *Journal of Human Ecology* 16(1): 33-42.

FSI, 2009. State Forest Report, 2009. Forest Survey of India, Government of India, Dehra dun.

Gadgil, M., Berkes, F. and Folke, C. 1993. Indigenous knowledge for Biodiversity Conservation. *Ambio* 22 (2-3): 151-156.

Malhotra, K.C., 1990. Village Supply and Safety Forest in Mizoram: A Traditional Practice of Protecting Ecosystem. Abstracts of V International Congress of Ecology, p. 439.

Satpathy, K. K., Sarma, B. K., Goswami, S. N., and Verma, N. D. 2003. *Developing Land Resources.* Ministry of Rural Development, Government of India, New Delhi.

Syiemlieh, H.J. 2007. Shifting cultivation in North-Eastern India-Some reflection. In: Saxena, K. G., Liang, L., Rerkasem, K (Eds) Shifting Agriculture in Asia: *Implications for Environmental Conservation and Sustainable Livelihood.* M/s Bishen Singh Mahendra Pal Singh, Dehra Dun, India.

Tiwari, B. K., Barik, S. K. and Tripathi, R. S. 1999. *Sacred forests of Meghalaya-Biological and Cultural Diversity*. Regional Centre, National Afforestation and Eco-Development Board, North eastern Hill University, Shillong.

Tiwari, B. K and Kumar, C. 2008. *Forest Products of Meghalaya: Present Status and Future Perspective.* Regional Centre, National Afforestation and Eco-Development Board, North eastern Hill University, Shillong.

Tiwari, B.K. 2005. Forest biodiversity management and livelihood enhancing practices of War Khasis of Meghalaya, India. In: Thomas,Y., Karki, M., Gurung, K., Parajuli, D. (Eds.), *Himalayan Medicinal and Aromatic Plants, Balancing Use and Conservation*. His Majesty's Government of Nepal, Ministry of Forest and Soil Conservation, Kathmandu, pp. 240-252.

Tynsong, H., Tiwari, B.K., Lynser, M. B. 2006. Medicinal plants of Meghalaya. *Medplant Network News*, 6, 7-10.

Tynsong, H. 2009. *Plant Diversity and NTFP management in the Community Forests of War Area, Meghalaya.* Unpublished Ph. D Thesis, Department of Environmental Studies, North eastern Hill University, Shillong.

3

Forest Economy and the Tribes of North-East India: Some Emerging Issues

S. B. Chakrabarti

The politico-jural transformation of the entire tract of the present day north-east India, as is known in the common frame of geographical reference, has a fairly long history. And much of it has been already discussed by scholars and administrators across the county and beyond. It is not necessary for us at the moment to enter into that discussion. The point of departure being the annexation of the area by the British rulers since 1826, it would be convenient only to mention that before Independence the districts of Assam were guided by the Chin-Hill Regulations of 1896 with respect to access to land and forest. These regulations recognised the rights of the village communities over their respective territories. This very principle and provision were subsequently incorporated in the Sixth Schedule to the Constitution of India in 1950[1]. One may only like to link up the background of this provision just mentioned above by bringing a reference form the monumental writings of a British administrator, namely Baden-Powell who mentioned this in his book of 1892 entitled *The Land Systems of British India*.[2]

While discussing about the hill districts of Assam Province (located in the centre of the Province, and also in the frontiers, subject to special rules) the author pointed out that those hills were "inhabited by various tribes, more or less civilized or barbarous, on the north and north east, as well as with the Central Hills of the Assam Range. On the south too, Sylhet and Cachar are in contact with the hills of the Lushai Country. Some of the tribes occupying the hills are independent, or in merely political relations with the Government; others are under British administration, but are not advanced enough to be under the same Civil, Criminal and Revenue laws as the older districts of the plains. It is, therefore, necessary (a) not only to provide a simple form of administration for such hill districts as are British, but also (b) in the case of the frontier and other hill tracts, to regulate the intercourse between the inhabitants of the plains and the hill tribes, whose country presents attractions in the shape of a trade in India - rubber and ivory. If the land holding in the hills and the trade intercourse were not regulated, complications and quarrels would be sure to ensue. In 1873, therefore, by Regulation V (of that year), a law was made, the object of which was to enable an 'inner line' to be drawn between the hill tribes and their neighbours in the plains. The holding of land beyond this line by strangers, and the intercourse for trade purposes or collecting forest product, is prohibited or regulated. The Regulation has ceased to apply to the Garo Hills, and no inner line has been found necessary in the Khasi Hills; but it is still in force on the northern frontier and to the south of Cachar…. Besides this inner line Regulation, the Regulation II of 1880, as extended by III of 1884, may be applied to all the hill districts directly under administration as British territory" (Baden-Powell, 1892, pp. 451-452).

The principles of 'seclusion' and 'protection' of the major territories belonging to the present day north east India have been experimentally evolved form time to time form pre-independence period till the initiation of Constitution of independent India. The 'Scheduled District' of 1874 to the Sixth Schedule of 1950 through the 'Backward Tracts' of the Govt. of India Act of 1919, and 'Excluded Areas' and 'Partially Excluded Areas' of the Government of India Act 1935 are now fairly documented.[3] We need not take up that historical context here for our present discussion. We may only mention about the kind of social formation and transformation that has taken place in this part of the country. This process has been succinctly summarized by Bhagabati (1988:2).[4]

"There is enough evidence to surmise that before the advent of British rule-which mainly began with annexation of Assam in 1826; tribal transformation was largely a consequence of migration. Adoption of viable strategies for adaptation to respective habitats by each group, and working out ways of dealing with other groups of people in a situation of contact. On the whole there was perhaps greater fluidity in the tribal situation so that many of the tribes were not as sharply defined social formation as they came to be in the subsequent years. In the case of Assam the Ahoms did not have any real sovereignty over the hills and their inhabitants. Bhattacharjee also pointed out that "The social and polity formation processes in North-East India were spontaneously influenced by its geo-political situation which absorbed pan-Indian traditions and developments in neighbouring South East Asia. In the past, the region experienced a fine blending of these two traditions represented by the Indo-Aryans and the Indo-Mongoloid tribal communities. The emergence fo states form the indigenous and immigrant tribal social bases was a significant development in the mediaeval history of the region" (1991:11).[5]

Tribals' age-old symbiotic relationship with forest is well known. The Report of the Committee on Forest and Tribals in India[6], which was prepared under the directive of the Ministry of Home affairs in the early 1980s noted that "this symbiotic relationship suffered a set back during the colonial rule when forest was looked upon only as a source of maximization of profit and not as a vital link between human habitat and the larger environment... There can not be any development of forests without development of the forest dwelling tribal communities ...The scheduled tribes live mostly in forest areas..... Therefore, the two directive principles of the Constitution, namely Article46 and Article48A, which seek to protect the economic interest of the forest tribes remain mutually reinforcing".[6]

Integrated development of forests and tribes have been the major concern right from the Dhebar Commission of 1961, the National Commission of Agriculture of 1976, Central Board of Forestry from time to time between 1950 to 1980, the National Forest Policy of 1988, the National Tribal Policy of 1990s to the Scheduled Tribes (Recognition of Forest Rights) Bill of the early 2005[7]. Significantly while the British Forest Policy of 1894 recognised the rights and privileges of the tribes, this became the rights and concessions later on. Subsequently, only concessions were provided to these people. The condition reverted back onto the question of granting rights to the forest dwelling tribes.

In view of the preceding observations it becomes imperative to critically review these vital policy documents and their implications in the context of defending the interest of the forest tribes. Forest in north east India has become extremely important on two major aspects, namely

(i) this region is considered as one of the great natural resource bases in the country which presents rich bio-diversities along with many rare medicinal plants[8], and (ii) this vast hilly terrain has appeared as the classical centre of pursuing the archaic pattern of jhum cultivation[9] on the one hand and its later improvements of modern agri-horticultural activities on the other[10]. In addition to that the emerging mining and quarrying exploitations are also on the increase in some parts of north-east India which have opened new space for manipulative politico-economic interface for human interaction beyond the boundaries of its immediate environ.[11]

The north-east India by and large has supported about 63 per cent of the county's green cover. Therefore, if unchecked depletion should legitimately attract public reaction and cannot simply go unnoticed. Multiple factors, both techno-economic[12] and socio-political[13] have been responsible for this ruin of the forest cover. These factors have their obvious roots in the local historical milieu as well as have other supra local connectivities[14]. The entire social spectrum is embedded in a mix between the traditional community control of land and forest[15] on the one hand and the emerging trend of privatization over the common property resource on the other.[16]

The north-east Indian states are naturally endowed with a variety of ecological characteristics. Juxtaposed with this a lot of very important socio-cultural and politico-economic dimensions have enriched its unique variables. In the backdrop of this scenario what becomes important for scholars and researchers is to generate a strong data base at the village level on each state[17]. In the event of the most land in more areas being surveyed and the primordial customary rights prevailing over them althrough, the techno-economic planners as well as the political decision makers remain incredibly bogged down with dilemma for taking up comprehensive plan for development[18]. The questions of customary right and community ownership over land and forest often confront with a normative journal rights of the state administration[19]. The evolved managerial middle class maneuvering in the slow process of privatization over the common property resource (especially land and forest) has been another crux of the problem[20]. These are the changing phases of social-economic transformations in the whole of north-east India at the present moment besides other techno-economic and socio-political constraints and increasing demographic and encroaching market pressures.[21]

In order to understand the present realities of man-forest interface in its proper perspective one needs to take a balanced view between the subjective codifications of the modern administration and the objective moral governance of the generative traditional social institutions either in a local Village Council or its higher form in an Autonomous District Council. The future course of social transformation has also to be studied in the context of the capabilities of the region for preparing its huge human resource for all-round development.[22]

References

L. G. Shullai N.D. *A Brief Note on the Sixth Schedule to the Constitution of India*. Govt. Press, Shillong.

B.H. Baden-Powell 1892 *The Land Systems of British India*. Delhi: Orient Publishers (reprinted in 1974).

Govt. of India Acts 1919 and 1935.

A. C. Bhagabati 1988 "*Tribal Transformation in Assam and North east India: an Appraisal of Emerging Ideological Dimension*", Presidential Address to Anthropology and Archaeology section, 75th Indian Science Congress, Pune.

J.B. Bhattacharjee 1991 *Social and Polity Formations in Pre-Colonial North east India*, New Delhi: Vikas Publishing House Pvt. Ltd.

Govt. of India, Ministry of Home Affairs 1982 Report of Committee on Forest and Tribals in India, New Delhi.

Reports of Govt. of India—1961, 1976,1988, 1999, 2005.

S. Mukhopadhyay 2003 "*Boi-diversity: Immense Source of Wealth and the Impact of its Loss on the Environment*", *Northeastern Himalayan biodiversity*, (Ed.)S. Chakravorti, Calcutta: Humboldt Club, pp 42-47.

NEISSR 1976 *Shifting Cultivation in Northeast India*, Shillong.

Atul Sarma 2002 "*Economic Development of the Northeastern States in the Context of Globalisation*", Contemporary India, Vol.1, No. 4, pp 91-111.

Patricia Mukhim 2009 Article in the Statesman,June,8.

Patricia Mukhim 2008 Article in the Statesman, November,17.

Patricia Mukhim 2009 Article in the Statesman, April 27.

Sajal Nag 2002 "*Structure of a Non-State Economy: Political Economy of Terrorism in North east India*", Contemporary India, Vol.1, No. 4, pp/ 189-212.

L. S. Gassah 1997 (Ed.) The Autonomous District Councils, New Delhi, Omsons Publications.

L. B. S. National Academy of Administration 1995 *Land Reforms: Agenda for North East.*

J.N.Das (et.el) 1989 *Land System of Arunachal Pradesh*, Bombay: N.M.Tripathi Pvt. Ltd.

H.C. Saratchandra (et.el) 1980 *Development Without Destruction*, Shillong (Papers presented in a symposium organized by the Forest dept.,govt. of Meghalaya).

B.B.Dutta and P.S. Datta 1986 "*Landholding Pattern Among the Khasi-Jaintias: the tradition and the deviation*", B.N.Bordoloi (Ed.) *Alienation of Tribal Land and Indebtedness*, Gauhati: tribal Research Institute, Govt. of Assam, pp 149-156.

Monalisa Chankija 2009 Article in The Statesman,January,26.

Jayanta Madhab 2002 "*Social Conflicts in the North east: An economic Interpretation*", Contemporary India, Vol.1, No. 4, pp 177-188.

Kamal Nandi (Ed.) 1982 Resource potentials of North east, Vol.I, Shillong: Meghalaya science Society and R. S. Tripathi (Ed.) 1984 Resource potentials of North east, Vol.II, Shillong: Meghalaya science Society.

4

Shifting Cultivation: One of the Root Causes of Deforestation in North-East India with Special Reference to Meghalaya

Sanjoy Choudhury

The forest ecosystem consists of major components like atmosphere, climate, soil and its living organisms. They influence the environmental quality of the earth. Forests play a major role in regularly socio-economic condition of a country besides controlling the ecological balance. Over the years forests in the country have suffered serious depletion. Forests are one of the most important natural resources of the north eastern region of India.

India is one of the few countries of the world with rich forests. It is interesting to note that while the country has lost 6.8 ha forests between 1972 and 1983,the forest cover from 1983 onwards in nearly stable. Compared to the rest of the country forests in the north –eastern hill region which forms a part of the Esatern Himalaya are undergoing deforestation at a rapid rate. This is primarily due to the increasing demand for land shifting cultivation(loacly called Jhumming). However, the latest survey indicated that more forests are put to jhumming than abandoned for post jhum regeneration annually of the eight states in the region, Mizoram, Meghalaya, Nagaland and A.P. facing maximum deforestation (Report by National Remote Sensing Agency).

The study attempts to monitor the affect t of shifting cultivation on forest cover in Meghalaya.

Shifting cultivation is prevalent in all the northeastern states. Studies carried out by the Forest Survey of India, i.e. State of Forest Report, 1995 and 1997 indicate that loss in forest cover in the northeastern states was mainly due to the shifting cultivation. In the State of Assam, other factors have also been shown to contribute to reduction in the forest cover.

The shifting cultivation is generally practised in the following sequence:

- Selecting a forest patch and clear fell the vegetation normally in December and January
- Burning of the vegetation. Small, cut-trunks portion and roots are normally not removed. The herbs, shrubs and twigs and branches (slashed vegetation) are burnt in February and March
- Sowing of seeds, by dibbling, generally of cereals, vegetables and oil seeds in April–May
- Continuing cultivation for a few years
- Abandoning the cultivated site and shifting to other forest sites
- Returning to the former site, and once again practise shifting cultivation on it.

With reduction in *jhum* cycle from 20–30 years to 2–3 years, the land under shifting cultivation looses its nutrients and the top soil. With reduction in crop yield, the families start moving to other virgin areas. Now a stage has come that it has already affected 2.7 million ha of land, and each year 0.45 ha of land fall under shifting cultivation, in north east India.

Frequent shifting from one land to the other has affected the ecology of these regions. The area under natural forest has declined; the fragmentation of habitat, local disappearance of native species and invasion by exotic weeds and other plants are some of the other ecological consequences of shifting agriculture. The area having *jhum* cycle of 5 and 10 years is more vulnerable to weed invasion compared to *jhum* cycle of 15 years. The area with fifteen-year *jhum* cycle has more soil nutrients, larger number of species, and higher agronomic yield with ratio of energy output to input as 25.6 compared to *jhum* cycle of 10 and 5 years (4.6–9.8) (ref. 10). Similarly, while studying *jhum* ecology in Meghalaya, it was reported that water and nutrient losses in shifting -cultivation areas were far greater than in the virgin areas, and areas left for 50 years after *Jhuming*. Thus, reduction in the cycles of *podu*, or jhuming, adversely affects the recovery of soil fertility, and the nutrient conservation by the ecosystem. Repeated short-cycle *jhuming* has created forest-canopy gaps which are evident from the barren hills.

Shifting Cultivation & its features with special reference to Meghalaya

Shifting cultivation commonly known as "*jhumming*" is one of the most ancient systems of farming believed to have originated in the Neolithic period around 7000 BC. It is called slash and burn method of cultivation. The system is regarded as the first step in transition from food gathering and hunting to crop husbandry. It is still practised in the hill areas of the North east states and in certain pockets of the country. Among the Khasis, this system of cultivation is known as "*Thang shyrti*" or "*Thang bun*". Essential features of *jhum* cultivation with special reference to Khasi Hills is as follows:

1. **Site selection**: Selection of sites for *jhum* practices is usually done by the individual owners of the land according to the needs of the family and family size. This is usually done before the month of December.
2. **Jungle cutting**: This operation is usually done in the months of December to January.
3. **Drying of Debris**: This operation follows just after the jungle is cleared and the debris is left to dry in the open.
4. **Burning**: Setting of fire to the dried debris is done from mid February to March. This operation is done with care so as to avoid forest fires from spreading across to other hills, which were not meant for '*jhumming*'.
5. **Sowing/Planting**: Sowing and planting of various crops is done in an intimate mixture by dibbling. Upland paddy is the main crop grown in mixture with maize, millet, sorghum, tapioca, chillies, cotton, turmeric, pumpkin, etc.
6. **Abandoning of Land** after 3-5 years of cultivation and shifted to another site for repeating the same process.

7. **The hutments** of the village now remain at the same place in contrast to earlier days when the whole village shifted to new site.
8. **Cropping is done with minimum tillage**: No animals or large implements are used by the *jhum* cultivators for preparing land. The only implements used are the chopping knife, sickle, dibbling stick, spade and hoe.
9. There is practically **no capital investment** except **labour** and the **seeds**, which usually come from the household.

Jhum cultivation has a chain of effects and after-effects leading to multifarious adverse conditions. The following chart shows, at a glance, how the system causes far-reaching disturbances in the conditions of the soil leading to changes in the climate, ecological imbalance and the environmental degradation, besides the immediate results of fertility loss and low productivity of crops.

Firstly, the process of cutting and burning, *jhum* causes denudation of forests leading to two consequences, viz. ecological imbalance and elimination of the sources of water. Secondly, it causes soil erosion leading to silting up of rivers and streams which causes flood, on the other hand it affects fertility of the soil resulting in low productivity and pressure on land. Thirdly, the system of *jhum* is a primitive method with low technology involving much labour without subsidiary income. Lastly, it is a socially degrading system, detrimental to development programme and conservative in spirit and attitude of the *jhumias*.

Present Status of *Jhum* Cultivation

The estimated *jhummia* population dependent on *jhum* is 13.87 per cent of the total rural population of 18.53 lakhs of the state in the year 2001. This clearly indicate that the population dependent on *jhum* had declined by 27.04 per cent in a period of thirty years time that is from 40.91 per cent out of the total rural population of 8.65 of the state in the year 1971.

At present with rapid increase in human population, there is corresponding heavy pressure on land. The *jhum* cycle has reduced drastically over the last few decades to an average of below ten years and in extreme cases such as that of Garo hills to as short as 3 to 5 years. In due course of time this cycle may even be brought down further. The soil has lost its sufficient period of recuperation and consequent capacity to regain its lost fertility and the immediate consequence has been the drastic reduction in the productivity of *jhum* system. This has resulted in increase food insecurity and the associated problems such as acute soil erosion and fast depletion of the valuable fertile topsoil, loss of flora and fauna, loss of reduced availability of water both for drinking and irrigation, siltation of lakes and rivers and consequent floods, loss of valuable forest species etc. In total it increases the destruction of environment and disturbing/upsetting the ecological balance of nature. This has far reaching affect and consequences to the welfare and prosperity of the inhabitants themselves.

It has been estimated that the average soil loss from the first year and second year of *jhumming* is 146.6 and 170.2 tonnes per hectare per year respectively. This clearly indicates that the second year of *jhumming* is comparatively more hazardous than the first year. Together with the loss of valuable fertile topsoil there is also corresponding great loss of nutrients available in the soil.

If we were to convert the above physical destruction and loss into monetary or financial terms, perhaps the figures will be stupendous — it may come to hundreds of thousands of crores of rupees. This is indeed a colossal loss to the people and the state. Against this background, the yield obtained from *jhum* fields at present is only a pittance and can sustain the *jhummia* families for only about four to five months in a year. This is indeed a picture of definitely unequal comparison in terms of economic development and loss of vital natural resources of nature.

Table: 1 – 1981

Name of District	Total Population (Rural)	*Jhumia* Families dependent on *Jhum*	*Jhumia* Population dependent on *Jhum*	Per centage of the *Jhumia* population from the total rural population	Annual Area under *Jhum* in sq.km.	Per centage of Annual *Jhum* Area from the total geographical area
East Khasi Hills Ri Bhoi	3,30,614	7,029	35,540	10.75	103.29	1.99
West Khasi Hills	1,57,696	3,629	19,207	12.18	53.33	1.02
Jaintia Hills	1,43,479	3,251	14,902	10.39	47.77	1.25
East Garo Hills	1,32,260	8,935	46,512	35.17	131.29	5.04
West Garo Hills South Garo Hills	3,30,437	28,877	1,41,786	42.91	424.32	7.63
Total	**10,94,486**	**51,721**	**2,57,947**	**23.57**	**760.00**	**3.39**

Table: 2-1991

<table>
<tr><th>Name of District</th><th>Total Population (Rural)</th><th>Jhumia Families dependent on Jhum</th><th>Jhumia Population dependent on Jhum</th><th>Per centage of the Jhumia population from the total rural population</th><th>Annual Area under Jhum in sq.km.</th><th>Per centage of Annual Jhum Area from the total geographical area</th></tr>
<tr><td>East Khasi Hills</td><td rowspan="2">4,34,075</td><td rowspan="2">10,811</td><td rowspan="2">54,055</td><td rowspan="2">12.45</td><td rowspan="2">110.26</td><td rowspan="2">2.12</td></tr>
<tr><td>Ri Bhoi</td></tr>
<tr><td>West Khasi Hills</td><td>2,05,818</td><td>5,085</td><td>25,425</td><td>12.35</td><td>51.86</td><td>0.99</td></tr>
<tr><td>Jaintia Hills</td><td>1,99,872</td><td>2,683</td><td>13,415</td><td>6.71</td><td>27.36</td><td>0.72</td></tr>
<tr><td>East Garo Hills</td><td>1,76,826</td><td>12,076</td><td>60,380</td><td>34.15</td><td>123.16</td><td>4.73</td></tr>
<tr><td>West Garo Hills</td><td rowspan="2">4,28,140</td><td rowspan="2">37,584</td><td rowspan="2">1,87,920</td><td rowspan="2">43.89</td><td rowspan="2">383.36</td><td rowspan="2">6.89</td></tr>
<tr><td>South Garo Hills</td></tr>
<tr><td>Total</td><td>14,44,731</td><td>68,239</td><td>3,41,195</td><td>23.62</td><td>696.00</td><td>3.10</td></tr>
</table>

Table: 3- 2001

<table>
<tr><th>Name of District</th><th>Total Population (Rural)</th><th>Jhumia Families depen-dent on Jhum</th><th>Jhumia Population Dependent on Jhum</th><th colspan="2">Per centage of the Jhumia population from the total rural population</th><th colspan="2">Annual Area under Jhum in sq.km.</th><th colspan="2">Per centage of Annual Jhum Area from the total geographical area</th></tr>
<tr><td>East Khasi Hills</td><td>3,83,027</td><td>1,721</td><td>3,605</td><td rowspan="2">4.51</td><td>0.94</td><td>6.20</td><td rowspan="2">43.6</td><td>0.23</td><td rowspan="2">0.84</td></tr>
<tr><td>Ri-Bhoi</td><td>1,79,630</td><td>4,351</td><td>21,755</td><td>12.11</td><td>37.40</td><td>1.53</td></tr>
<tr><td>West Khasi Hills</td><td>2,60,595</td><td>5,374</td><td>26,870</td><td colspan="2">10.31</td><td colspan="2">46.19</td><td colspan="2">0.88</td></tr>
<tr><td>Jaintia Hills</td><td>2,70,669</td><td>1,366</td><td>6,830</td><td colspan="2">2.52</td><td colspan="2">11.74</td><td colspan="2">0.31</td></tr>
<tr><td>East Garo Hills</td><td>2,11,652</td><td>13,630</td><td>68,150</td><td colspan="2">32.20</td><td colspan="2">117.15</td><td colspan="2">4.50</td></tr>
<tr><td>West Garo Hills</td><td>4,57,422</td><td>18,086</td><td>90,430</td><td rowspan="2">23.71</td><td>19.77</td><td rowspan="2">155.45</td><td rowspan="2">223.32</td><td>4.19</td><td rowspan="2">4.01</td></tr>
<tr><td>South Garo Hills</td><td>90,462</td><td>7,900</td><td>39,500</td><td>43.66</td><td>3.67</td></tr>
<tr><td>Total</td><td>18,53,457</td><td>42,428</td><td>2,57,140</td><td></td><td>13.87</td><td></td><td>442.00</td><td></td><td>1.97</td></tr>
</table>

Table: 4

Year	Total Rural Population (in lakhs)	Increased population (per cent)	*Jhumia* Families dependent on *Jhum* (in Nos.)	Decline/ Increase of *jhumia* families (per cent) Decline=(-) Increase=(+)	*Jhumia* population (in Nos.)	Per centage of *Jhumia* population from the total rural population	Annual Area under *Jhum* (in square kilo-meters.)	Per centage of Annual *Jhum* Area from the total geographical Area
1971	8.65	-	70,000	-	3,53,850	40.91		-
1981	10.94	26.47	51,720	(-)26.11	2,57,947	23.57	760	3.39
1991	14.45	32.08	68,239	(+)31.94	3,41,195	23.62	696	3.10
2001	18.53	28.23	51,428	(-)24.60	2,57,140	13.87	442	1.97

Underlying Factors Contributing to the Persistency of *Jhum* Cultivation in Meghalaya

1. **Socio-Cultural and Religious Significance**: The people who practice *Jhum* cultivation do so because their ancestors had practiced it from centuries, much before scientific agriculture and forests management was thought of and because their ancestors had thought it as the best way of living. The shifting cultivators generally care for the forest and some even worship them. Therefore, while the system for site selection and rituals adopted may vary from tribe to tribe or region-to-region, yet all of them believe in a good burn for a bumper crop.
2. **Earnest and Cheapest Method**: The method involves minimal land preparation insignificant capital investment, so manures or fertilizer requirement and manual labour is the only energy input, which is largely contributed by family members.
3. **Absence of Land Ownership**: As the land belongs to the community and in the absence of a secure tenure, the *jhummias* are unwilling to invest in land development activities. This lack of ownership makes the *jhummia* unable to have access to bank facilities. However, with increasing communication of life and with certain community land

being released for sale, there is also every likelihood of the danger of usurpation of land by the affluent of society and consequential marginalisation of poor farmers of *jhummias* in rural areas.

4. **Mixed Cropping**: For the sustenance of the family, the *jhummia* needs the various crops that he can grows in the *jhum* plot whereas in the wet terraces only one crop can be grown. Also, the presence of different crops in the same plot is considered good for pest management due to genetic diversity and the sequential harvesting of crops in an effective way of managing many species over both space and time and contributes to agro-ecosystem stability, besides showing better orientation of nutrient use efficiency.

5. **Lack of Alternatives**: In many areas of the State, the *jhummias* are becoming aware of the ill-effects of shifting cultivation such as dwindling productivity, labour intensiveness, poor living conditions, deteriorating environment, etc. With the increasing pressure of population, expanding commercialisation of agriculture, acquisition of land by affluent and such among tribals as future investment potential and exploitation of timber, there is a serious degradation of the environment, which is threatening the carrying capacity of nature. The *jhummias*, though are conscious of the monetised economy and would like to respond to the interventions in the form of inputs and marketing, yet are unable to do so due to the lack of location specific and need based programmes and solutions such as allotment of wetland terraces into assured irrigation particularly in paddy growing tribes.

6. **Lack of Improved Technology Awareness**: In a State where ethnicity is prominent and there is aversion to bring labour from outside to cultivate the land, the continuance of shifting cultivation practice would largely depend on the kind of developmental activity that is being offered. The lack of creating awareness and non-introduction of improved technology that are location specific and need based would only strengthen the farmers resolve to resist change and persist with *jhum*.

Therefore, creation of awareness through exposure of shifting cultivators to markets, valley cultivation, Government agencies and strengthening and redesigning of agro-factory systems, animal husbandry practices etc. by incorporating traditional knowledge along with the introduction of appropriate rural technology such as bio-gas, rain water harvesting and artisan skill development will lead to learning and expanding their options for a sustainable rural livelihood.

7. **Lack of Financial Support**: Besides the bottlenecks of development highlighted above, one of the biggest constraints in livelihood matters relating to *jhum* control, poverty alienation, conservation oriented and other income generating programmes is the lack of extension of credit. Traditional land tenure is blessed as constraining security for the credit. However, this appears to be only an excuse for non-performance and reluctance for credit extension. Therefore, with this in mind, the *jhummias* has no alternative but to continue with his age-old practice of *jhum*.

8. **Lack of Transport and Communication Linkages**: The long isolation of *jhummias* by such constraints as terrain and topography, deficiency in infrastructural support, historical isolation, and unbalanced economic growth and inadequate delivery of social services such as health, education, etc., must end if they are to look for other rural livelihood option besides shifting cultivation. The severance of their natural markets across the international borders has also uniquely disadvantaged the *jhummias* from taking up more productive and competitive agriculture.

9. **Lack of Co-ordinated Multi-Disciplinary Approach**: The *jhum* control programmes are being implemented by different developmental agencies in relative isolation without proper co-ordination through multi-disciplinary approach. This lack of institutional consultations leads to mainly personality-based consultation with each trying to out do or overdo others resulting in alienation of traditional systems and knowledge from the decision making and governing process, leading to marginalisation and non-participation by the *jhummias* in the developmental works.

10. **Poor Literacy**: It is a well-known fact that majority of the rural people are illiterate and generally ignorant of the various scientific and technological developments that is taking place around them, mainly due to the continuing isolation caused by geographical and other factors. In addition, the shortage of proper schools and colleges in *jhum* prone areas have left little scope for the *jhummias* to find out and learn about other alternatives to *jhum* cultivation.

Efficacy and Impact of Existing Programmes

Shifting cultivation, logging and fuel-wood harvesting are considered to be the chief reasons for environmental degradation. And with the increasing pressure on population, expanding commercialisation of agriculture, acquisition of land by the affluent and rich among the tribals as future investment potential, the *jhum* cycle has reduced considerably. This has resulted in large-scale erosion, drying up of rivers and stream, siltation of tanks and lakes and the receding of flora and fauna

Investment of *Jhum* Control Programmes (Schemes of the Government)

Recognizing the major problems on *jhum* cultivation in the State, the Government through the Soil and Water Conservation Department took up the scheme of *jhum* control in 1974-75 under the State Plan. This scheme was undertaken to provide effective supporting base for permanent settlement of the *jhummia* families in order to wean them away from *jhumming*. The scheme is a package programme and consists of the following components:

1. To provide permanent cultivation land (Bench Terraces) with assured irrigation facilities.
2. Follow up programmes by way of supply of inputs like seeds and plants, manure and fertilisers including cultivation cost for three years at a gradual sliding scale.
3. To provide cash/horticultural crop plantations and afforestation.
4. To provide drinking water supply for village settlement.
5. To provide link/approach roads to work areas.

Land development in the form of bench terraces with assured irrigation facilities have proved most effective means in attracting the *jhummias* to settled agriculture. Cash/Horticulture crop plantations undertaken under this scheme is also making an appreciable impact in the process of weaning away the *jhummias* from their age-old practice of *jhumming* and it could be seen in the field that most of the abandoned *jhum* areas have been converted into permanent cash/horticulture plantations such as arecanut, cashew, citrus, etc.

Table: 5

Cumulating Physical and Financial Achievement under State Plan *Jhum* Control Scheme Since (1974-75 to 2001-02)

Sl. No.	Activities	Physical Achievement upto 2001-2002	Financial Achievement upto 2001-2002 (Rs. in lakhs)	Total
1	Terracing a reclamation	8772.83 ha.	551.16	2965.95
2	Follow-up programme	5828.11 ha.	82.04	
3	Afforestation	3612.12 ha.	111.51	
4	Irrigation	6835.65 ha.	371.54	
5	Camp & camp equipments	445 Nos.	43.06	
6	Dwelling houses	1665 Nos.	33.31	
7	Drinking water	306 Nos.	25.31	
8	Link roads	669.69 km.	121.85	
9	Cash/horticulture crop development works	7244.59 ha.	1626.17	
10	Families benefited		8773 Nos.	

hum Cultivation Area of Meghalaya

- Block-wise Jhum Cultivation area – Not Available.
- District-wise Jhum Cultivation area as shown in the following table (based on the sample study conducted by the Task Force Constitute by the State Government to study the problems of Jhum cultivation in the State):

Table: 6

Sl. No.	Name of District	Annual Area under *Jhum* in Km^2 (Current year 2001)
1	East Khasi Hills	4.90
2	West Khasi Hills	44.6
3	Ri-Bhoi District	47.86
4	Jaintia Hills	10.25
5	East Garo Hills	92.61
6	West Garo Hills	144.69
7	South Garo Hills	62.41
Total		407.32

Table: 7

Sl. No.	Name of District	Annual Area affected by the *Jhum* for the last five years in Km^2
1	East Khasi Hills	24.50
2	West Khasi Hills	223.00
3	Ri-Bhoi District	239.30
4	Jaintia Hills	51.25
5	East Garo Hills	463.05
6	West Garo Hills	723.45
7	South Garo Hills	312.05
	Total	**2036.60 Km^2**

Bun Cultivation in Meghalaya

Shifting cultivation or *Jhumming* in Meghalaya is broadly classified into two categories based on the nature of the cultural operations: (1) Clear Felling *Jhumming* with least disturbance to the soil and (2) '*Bun*' Cultivation *Jhumming* involving disturbance to the soil. The second system of *jhumming*, locally known as '*Bun*' Cultivation, is mostly practiced by the *Jhummia* families of East Khasi Hills District, West Khasi Hills District and to some extent in Ri-Bhoi District and Jaintia Hills District.

In this system of *Jhumming*, the *jhummias* cut the shrubs and grasses (in the absence of forest cover) during the months of November/December and the cut materials are then laid in rows along the slope in bunds locally known as '*Bun*'. The soil surrounding the bunds/*Buns*, are then hoed and covered on top of these bunds/*Buns* during the months of December/ January. Burning is done during January and planting is carried out from the end of February till March depending on the type of crops grown.

In this type of *jhumming*, the soil is very much disturbed and as a result, heavy soil erosion occurs during the monsoon periods leading to the depletion of soil fertility and exposing the sub-soil strata. The *jhummia* families are then left with no option but to apply manures and fertilizers to the crops grown on these bunds/*buns*.

Meghalaya with a total geographical area of 22,429 square kilometerss. has a total of 2,64,960 number of families residing in 3,610 villages. As per the survey report conducted, the total number of *jhummia* families practicing *Bun* cultivation had been estimated to be at 10,262.

The District-wise Distribution of the *Jhummia* families is as follows:

Table: 8

Sl. No.	Districts	Area in Sq.kilometers.	No. of *Jhummias* Families	Families Practicing *Bun* Cultivation
1	East Khasi Hills	2748	56,718	9355
2	West Khasi Hills	5247	35,468	357
3	Jaintia Hills	3819	34,425	Nil
4	Ri-Bhoi	2448	24,220	556
5	East Garo Hills	2603	33,289	Nil
6	West Garo Hills	3714	59,386	Nil
7	South Garo Hills	1850	21,457	Nil

East Khasi Hills

***Bun* cultivation** is widely prevalent in the East Khasi Hills District. Majority of the farmers particularly the Mylliem C.D. Block, Mawphlang C.D. Block, Mawryngkneng C.D. Block and Mawkynrew C.D. Block, practice this type of cultivation. Potato, ginger, paddy, millet, maize, etc., are sown by the farmers as pure/mixed crop. Cultivation on the same plot is extended upto two years only and the plot is left fallow for a period of 3-5 years. Most of the farmers do not own the land but they cultivate the land on a long lease by paying an annual rent to the owners. Hence, it becomes a problem to take up any conservation measures as sometimes the lessee might agree but the owner refuses and vice-versa. It had been estimated that about 13,755 families in the whole East Khasi Hills District practiced '*Bun*' cultivation.

West Khasi Hills District

In West Khasi Hills, the number of families involved in '***Bun*** cultivation had been estimated at 551.Jaintia Hills District: The *jhummia* families involved in '*Bun*' cultivation had been estimated at 756.

Garo Hills

In the entire Garo Hills, there had been no practice of '*Bun*' cultivation.

Dependence on '*Bun*' Cultivation

Table: 9- District-wise Information of Villages and Families

Sl. No.	Name of District	Names of CD Blocks	Total No. of Villages	Total No. of Families	*Jhummia* Families practicing '*Bun*' Cultivation	Total Population	*Jhum* area per Family in ha.	Persons per Family
1	East Khasi Hills	1. Mylliem	97	13,150	3,075	70,607	1.00	5
		2. Mawphlang	158	7,873	5,590	46,685	1.50	6
		3. Mawryngkneng	62	6,326	2,564	35,914	0.50	6
		4. Mawkynrew	67	5,110	2,526	30,170	0.41	6
		5. Shella Bholaganj	191	7,736	Negligible	38,022	-	5
		6. Pynursla	152	9,404	Negligible	47,171	-	5
		7. Mawsynram	159	7,129	Negligible	38,194	-	5
		Total	**886**	**56,728**	**13,785**	**3,06,763**	**3.41**	**38**
2	West Khasi Hill	1. Nongstoin & Ranikor	240	7,852	Negligible	46,704	-	6
		2. Mairang	115	10,755	551	63,095	0.27	6
		3. Mawshynrut	261	7,069	Negligible	40,248	-	6
		4. Mawkyrwat	200	9,792	Negligible	55,771	-	6
Total			**816**	**35,468**	**551**	**2,05,818**	**0.27**	**24**
3	Ri-Bhoi	1. Umling	264	9,717	Negligible	50,561	-	5
		2. Mawphlang	296	14,503	756	76,751	0.75	5
		Total	260	24,220	756	1,27,312	0.75	10

Impact of Shifting Cultivation on Soil

Shifting cultivation is practiced in about 386.9 thousand ha annually and on an average 50 to 80 tones of dry matter is burnt per ha, depending on the shifting cycle. About 1.868, 21.511 and 2.228 million tones of leaves, food and litter containing 137.35, 7.46, 95.95 and 97.50 thousand tones of N P K and calcium, respectively, are burnt every year causing enormous loss of vegetation. Besides, creating, atmospheric pollution, it causes ecological and soil degration in the region. The shifting cultivation practice prevalent in the region causes tremendous loss of soil and nutrients

(Shahlace et al, 1991) whereas, these losses can be minimized to almost negligible level by managing the watersheds (Blackburn et al, 1986, 1990; Singh and Singh 1981, Satapathi, 1996). Table 1 shows the tremendous losses of soil and nutrients due to shifting cultivation. Annual soil loss has been estimated to be 88346 thousand tones and combined loss of N, P and K 17.092 thousand tones. The area under shifting cultivation varies from 2.6per cent in Assam to 96.0per cent in the Mizoram state of the net sown area (Sharma and Prasad, 1994).

Table 10:
Soil and available nutrients (N, P and K) loss due to shifting cultivation

State	Area under shifting cultivation ('000 km^2)	per cent of net sown area	Loss ('000 tonnes per year)	
			Soil	N+P+K
Arunachal Pradesh	70.0	4.7	14490	3.111
Assam	69.6	2.6	12318	1.685
Manipur	90.0	64.2	20430	4.289
Meghalaya	53.0	27.0	14151	2.999
Mizoram	63.0	46.0	13041	2.738
Nagaland	19.0	34.5	7962	1.432
Tripura	22.3	8.2	5954	0.839

Forest production on acid soils that are poor in basic *Catinos* is limited by mineral deficiencies. The sustained production of forest eco-system is closely linked to their nutrient cycle particularly in the tropical regions where greater per centage of nutrients is contained in the biomass. The burning of forest residues during land clearing leads to considerable nutrient loss from the eco-system (Toky and Ramkrishna, 1981). Strained forest productivity in acid soils requires the use of techniques which reduce nutrient export, increase nutrient accumulation in the biomass, increase efficiency of nutrient absorption and utilization and conserve water in the system. The hilly areas of North eastern states are becoming increasingly deforested and denuded due to over- exploitation and shifting cultivation. Biotic and abiotic interferences have caused considerable degradation to natural resources such as soil, water and forest to the extent that flow of rivers becomes destructive during flooding and insufficient during dry seasons. Soil conservation measure and afforestation play a definite role in reducing runoff and control of damage caused by excessive runoff. Vegetation has a great effect on soil loss due to runoff. Due to continuous deforestation in NE region, the rainfall in the region is showing a declining trend during the last 10 years.

Depletion of forest resources

Total forest area of North-Eastern Region is 167.4 thousand km², out of which 91.2 thousand km² (54.4per cent) area is dense forest cover and 76.2 thousand ha (45.6per cent) are open forests (Table 2). Arunachal Pradesh leads in forest cover with 68.6 thousand ha or 82.0per cent of its total geographical area. Assam has 24.0 thousand ha under forests but it is only 30.6per cent of its geographical area. As a whole, the region has 63.9per cent of its geographical area under the forest, which is well over the guidelines of the Government that in hills 60per cent area should be under forest

Table: 11: Forest area in different states of the region ('000 km²)

State	Geographical area	Dense forest	Open forest	Total forest	Forest area as per cent of geographical area
Arunachal Pradesh	83.6	54.2 (65.2)	14.4 (21.1)	68.6	82.0
Assam	78.4	15.7 (65.2)	8.3 (34.8)	24.0	30.6
Manipur	22.3	5.3 (30.2)	12.2 (69.8)	17.5	78.6
Meghalaya	22.4	4.0 (23.0)	11.7 (74.3)	15.7	70.0
Mizoram	21.1	4.3 (23.0)	14.3 (77.0)	18.6	88.1
Nagaland	16.6	3.5 (24.4)	10.9 (75.6)	14.4	86.2
Sikim	7.1	2.4 (77.5)	0.7 (22.5)	3.1	44.0
Tripura	10.5	1.8 (32.8)	3.7 (67.2)	5.5	52.8
Total	262.0	91.2 (54.4)	76.2 (45.6)	167.4	63.9

Source: The Hindu Survey of Environment, 1997.

Table: 12: Decline in forest cover between 1991 and 1995 in NE states

State	Forest cover (km^2)			Annual decrease (per cent)		
	1991	1993	1995	1991-93	1993-95	1991-95
Arunachal Pradesh	68750	68661	68621	0.069	0.029	0.049
Assam	24750	24508	24061	0.494	0.912	0.696
Manipur	17685	17621	17558	0.181	0.178	0.179
Meghalaya	15875	15769	15714	0.333	0.174	0.250
Mizoram	18853	18697	18576	0.413	0.323	0.367
Nagaland	14321	14348	14291	0.164	0.198	0.052
Sikkim	3127	3127	3119	-	0.128	0.064
Tripura	5535	5538	5538	+0.027	-	0.013
Total	168904	168269	167478	0.188	0.235	0.211

While 635 km^2 of forest area was lost between 1991 and 1993, 791 ha area depleted between 1993 and 1995. The largest per cent decrease in forest area was in Assam with 0.494per cent decrease during 1991 to 1993 and 0.921per cent decrease during 1993-95. Over all decrease in the region between 1991 and 1995 was 0.211 per cent. Recently, a news item (Sentinel, Feb., 24, 1998) indicated that the Government has shown concern over the alarming rate of decrease in forest area in the country that was 17,000 km^2 between 1995-1997. With the increase in the population and at the present Annual Compound Decrease Rate in forest area in the region, the forest cover would vanish within about next 100 years. The decline would follow exponential pattern, if care is not taken to preserve this valuable resource in the region.

Classifying shifting agriculture

- Professor B.K. Tiwari at North-Eastern Hill University, Shillong, classifies shifting agriculture into four types: traditional, distorted, innovated and modified. This classification aids in understanding how jhum has changed over the years and what is going wrong with newer practices of jhum.
- Professor Tiwari has mentioned that the emphasis should be on controlling distortions or retrogressive developments rather than on controlling shifting agriculture itself.

Discussion

It is commonly believed that jhum was a sustainable system in the past when fallow periods were long, but with increasing population pressure jhum cycles have declined and become unsustainable. The evidence for this is weak. In a review, Singh10 has pointed out that: (i) there is often no strong relationship between population pressure and jhum cycles, (ii) villagers choose to cultivate at cycles of 5-10 years even when longer fallow periods are possible, and (iii) that population density would impinge on jhum cycle only after some critical threshold of high population pressure is crossed. The duration of the fallow period is influenced by the ease of clearing the vegetation and soil fertility levels following the slash-and-burn operations. The burning of slash returns nutrients to the soil through ash and kills microbes allowing relatively high yields. Yields decline as the soil is depleted through one year of cropping and cultivation is rarely carried out for more than a year. When fields are abandoned, there is rapid regeneration of bamboo and other plants. After ten years, the vegetation and soil properties recover to levels that can support another round of jhum cultivation.11 In some areas with better regrowth of bamboo, even shorter fallow cycles may be feasible and sustainable.12 Where fallow cycles are less than five years, jhum cultivation may be stabilized using modern technical inputs, 13 while concurrently developing alternative village-based occupations.

The superiority of jhum cultivation over some forms of sedentary cultivation partly explains the persistence of this form of agriculture in North east India. Other reasons include the economic security provided by jhum and its cultural importance to indigenous tribes. Poor access to markets, capital, and technical know-how of more commercially rewarding alternatives such as horticulture and cash crop cultivation also hinders the transition to other occupations. Clearly, one cannot do away with jhum assuming it to be a primitive and inefficient system, as attempted in governmental *jhum* control programs and new land use policies.15 Instead, an unbiased understanding of the advantages of jhum is required for proper design and implementation of developmental programs. Erosion of valuable topsoil in the hills due to jhum has been alleged to cause siltation and floods in the plains. Singh16 has reviewed studies carried out by the Indian Council of Agricultural Research that compared soil erosion from jhum fields with other forms of cultivation on terraces and contour bunds. These studies show that jhum fields cultivated for a single year and abandoned (the most common practice) have less erosive losses of soil than the other forms of settled cultivation.Some erosion of topsoil is inevitable in any form of cultivation in the high rainfall hill tracts. Soil erosion is minimized in jhum due to the retaining of rootstocks of bamboo and trees in burned plots, the rapid recovery of weeds and bamboo following abandonment, and the interspersion of forests and fields on hill slopes.17 The evidence for siltation of rivers and floods because of soil erosion due to jhum is weak and possibly untenable. Other factors, such as large scale logging for timber extraction, may be responsible to a greater extent for the deforestation and environmental problems in North east India. To tackle these problems, the Government has tried various schemes to wean away *jhummias* in the last 30 years mainly through the introduction of plantation crops such as rubber, coffee, tea, black pepper, cashewnut, etc., through the line Departments.

Controlling measures

There has also been afforestation programmes, programmes of the Agriculture Department, centrally sponsored schemes and pilot project for controlling shifting cultivation. Schemes for land development and terracing with assured irrigation were also taken up to assist farmers for an alternative settled cultivation. These recently introduced schemes were taken up under the National Watershed Development Programmes in Rain-fed Areas (NWDPRA) and National Watershed Development Programme in Shifting Cultivation Areas (NWDPSCA) under central assistance .

Many *jhummias* have accepted these schemes and it is seen that the creation of wetland terraces with assured irrigation has attracted the *jhummias* particularly in the paddy growing areas. And with the *jhummias* becoming conscious of the monetised economy, the introduction of plantation/cash crops that assure perennial returns has had tremendous response for which there is not enough funds to meet the demands.

However, the poor extension services, lack of dedicated workers to save rural areas and the failure to ensure effective marketing linkages along with communication and transport network meant that the *jhummias* continue to miss the opportunities and remain isolated and backward. Besides the above, there are other factors that contribute to the overall stagnation and slow progress of the *jhummias* in our State. These are:

1. Lack of ownership right over the land.

2. Lack of inputs, improved technology and result oriented approach in interior areas.

3. Lack of integrated, co-ordinated and sustainable approach based in research and development, which will give a good impact to change the mindset of the *jhummias*.

4. Lack of proper co-ordinated efforts and effective linkages between research agencies and developmental agencies.

5. Lack of need based technology and flexibility of the scheme, particularly schemes from the Government of India.

6. Among poorer sections of the society, the long gestation period of plantation crops became a hindrance.

7. Lack of pre-implementation capacity building and farmers participation in the programme.

8. There is no continuity of Research in the *jhum* fields.)

In April 2006, the Government of Meghalaya agreed that it would no longer try to suppress shifting cultivation and would instead examine ways of integrating soil and water conservation measures within it. The National Forest Commission has included in its advice to the Government of India ICIMOD's recommendations stating that security of land tenure for shifting cultivators for both the agricultural and fallow phases should be increased. This can be done by reconsidering the classification of shifting cultivation areas and categorizing the as agricultural land with adaptive forest management in the fallow period.

A meaningful solution to the problem of Jhum has become critical ,not only from the point of biodiversity conservation, but also for productive agriculture in the region. The solution to the problem of the shortening fallow cycle lies in strengthening this already weakened agro-forestry system, through fallow management using appropriate tree species, rather than imposing an alien technology from outside the region.

A variety of other opportunities exist to redevelop land use systems other than jhum, such as valley wet rice cultivation, through appropriate designed traditional water harvesting and soil fertility management technologies. The strengthening of this land use could be also seen as a means to take away the pressure from jhum itself, to the extent possible.

References

Annual Report 1997-2008, ICAR Research Complex for N.E.H. Region.

Borthakur, D. N. Improving Productivity of Jhum cultivation with special reference to the North eastern Hill Regions. Indian Science Congress Association, 3rd Annual Session Waltair, January 3-7,1976.

Borthakur, D. N. Awasthi, R. P. and Ghosh, S. P. Alternative system of farming for increasing productivity in jhum lands. Proceedings of Seminar on Shifting Cultivation in N. E. India. North east India Council for Social Science Resaerch,1976.

Borthakur, D., N. Singh, A., Prasad, R. N. Shifting Cultivation in North east India and a Strategy for Land Water Resource Management: Beitrage trop. Land wirtsch Viterinarmed:23 H.2. 147-158,1985.

Down to Earth, Science and Environment Fortnightly, May 31st 2001.

Geography and You, Monthly Magazine on Development & Environment in India, Vol. 2, No.8 & 9,2002.

ICAR Research Complex, Shifting Cultivation in North East India: Monograph No. 3,1984.

Kushwaha, S. P. S., Hildebrandt, G. Shifting Cultivation as a Cause for Deforestation A Case of North-Eastern India.

Mawthoh, P. R. Shifting Cultivation in Meghalaya : paper presented at the Seminar on Shifting Cultivation in the North eastern States by N.E.C., May, 1989.

Rajiv Ranjan and V. S. Upadhyay, Ecological Problems due to Shifting Cultivation.

Singh, A. Shifting cultivation and Soil Erosion Problems in North eastern Hill Region. Paper presented in XVI Annual Convention of ISAE held at IIT, Kharagpur,1978.

Shifting Cultivation in North East India, North east India Council for Social Science Research.

State of Forest Report 2001-05, Forest Survey of India, Ministry of Environment and Forest Dehra Dun.

Siddheshwar Sarma, Meghalaya: The Land and Forest, A Remote Sensing Based Study, 2003.

Rathore, S. S., Choudhary, Geological Problems due to Shifting Cultivation.

5

Forest-Tribal Interface in North East India: The Road Ahead

Lalram Thanga

In recent years, the reform of the forestry sector has been debated and discussed at great length especially in States where externally-aided forestry projects are being operated. Several studies on the functioning of the sector have been commissioned by the donor agencies and a number of recommendations made on the reform needs of the sector. Even for north east India, it is indeed heartening to note that during the last few years there has been a significant rise in the availability of authentic reports based on studies concerning the forests and tribals. The question now is 'Where do we go from here and how do we convince the people in the corridors of power at the States and Centre to bring about this change?'

One size doesn't fit all

The country has been trying to attend to the developmental and welfare concerns of this multi-ethnic nation through a host of Centrally Sponsored Schemes of the Central Government. However, more often than not, the guidelines are tailor-made for the peninsular side where the majority resides.

The north-east, quite unique socio-culturally and whose connection with mainland India is only via the chicken-neck of a corridor somewhere near Siliguri has to also tow the line in order to avail the largesse.

Realizing this in recent times, the Central government has also started giving flexibility in a number of the schemes to suit local conditions. This, however, is still far from satisfactory. The present predicament of the Centre is that giving too much flexibility will result in improper implementation. However, this can be addressed through proper monitoring and evaluation. Even the RTI Act is also an important tool in this.

At the same time, the Centre has to note that north east region should not and cannot be treated as a single entity. The region is again made up of different ethnic groups having different cultures, customs and traditions. Hence, one size does not fit all!

Tribal Development:

The problems of the tribals and other backward communities have been drawing the attention of the government right from the framers of the Constitution although the achievement in this endeavour has been far from satisfactory. This is being addressed again in the Eleventh Plan (2007-2012) document of the Government of India, which stated thus:

"The Eleventh Plan will attempt a paradigm shift with respect to the overall empowerment of the tribal people, keeping the issues related to governance at the Centre. The operational imperatives of the Fifth Schedule, TSP 1976, PESA 1996, RFRA 2006; the desirability of a tribal-centric, tribal-participative and tribal-managed development process; and the need for a conscious departure from dependence on a largely under-effective official delivery system will be kept in view during this shift".

It may be recalled that a separate Ministry of Tribal Affairs exclusively meant for development of the tribals has also been established since 1999 and dedicated funds continued under the Tribal Sub-Plan (TSP).

The National Forest Policy 1988, has also made a specific clause for addressing the problems of tribals –

"*The life of tribals and other poor living within and near forests revolves around forests. The rights and concessions enjoyed by them should be fully protected.*"

In pursuance of this policy, Government of India has sent many directives including the comprehensive guidelines on tribal-forest interface in 1990 to deal with regularization of encroachments on forest lands, review of disputed claims over forest land arising out of forest settlement, disputes regarding pattas/leases/grants involving forest land – settlement thereof, elimination of intermediaries and payment of fair wages to the labourers on forestry works, conversion of forest villages into revenue villages and settlement of other old habitations and payment of compensation for loss of life and property due to predation – depredation by wild animals.Again in the same year, the Joint Forest Management guidelines were sent to all the States.

Ironically though, the guidelines on tribal-forest interface were not followed both in letter and spirit by the States thus resulting in the present imbroglio which ultimately led to legislation of the Scheduled Tribes & Other Forest Dwellers (Recognition of Forest Rights) Act 2006.

Recognition of Forest Rights Act *vis-a-vis* NE Region

The Forest Rights Act basically seeks to legalise existing rights to undo a historic injustice against indigenous and tribal peoples of India. It is applicable to all the States/UTs' excluding J&K. However, there are certain peculiarities of the north eastern states which have relatively more autonomy in the disposal of their affairs.

For example:

- **Nagaland and Mizoram: No effect unless the State Assembly passes resolution** - Article 371 (A) & (G) provides that "(1) Notwithstanding anything in this Constitution,—(a) no Act of Parliament in respect of—(ii) Naga customary law and procedure (iv) ownership and transfer of land and its resources, shall apply to the State of Nagaland/ Mizoram unless the Legislative Assembly of Nagaland & Mizoram by a resolution so decides."
- **Areas under the Sixth Schedule** -The areas of Assam, Tripura, Mizoram and Meghalaya that fall under the Sixth Schedule have a different type of protection. The Sixth Schedule clearly provides that the Autonomous District Councils (ADCs) of these areas have the power to make laws with respect to forests (excepting reserved forests) and with respect to land.

Communities still retain control over much of the region's natural forest ecosystems: either through the District Council (as in the case of Meghalaya, Mizoram, Tripura and the Karbi–Anglong district of Assam) or within the control of the clan, village or tribe (as in the case of Nagaland or Arunachal Pradesh. The district councils head the autonomous districts. District Councils are autonomous bodies that were created in order to balance the need for special protection and administrative responsiveness to local needs with the national interests (Sixth Schedule).

However, despite these provisions, that all is not hunky-dory is again reflected in the recent National Environment Policy 2006 which stated thus, ".......while since antiquity forest dwelling communities had generally recognized traditional community entitlements over the forests, on account of which they had strong incentives to use the forests sustainably and to protect them from encroachers, following the commencement of formal forest laws and institutions in 1865, these entitlements were effectively extinguished in many parts of the country. Such disempowerment has led to the forests becoming open access in nature, leading to their gradual degradation in a classic manifestation of the '*Tragedy of the Commons*'. It has also led to perennial conflict between the forest dependent communities and the Forest Department, constituting a major denial of justice….."

Again, the unique nature of the forests of north east India is brought out in the Report of the National Forest Commission 2006 stating that "……in view of the different traditional land tenures and ownership rights, the Commission recommends that the traditional rights of the north-eastern people's forest and land must be honoured. They should have the right to conserve, manage and rationally utilize their forest in an ecologically sustainable manner."

Challenges and Problems in the implementation of the Centrally Sponsored Scheme Watershed Development Project in Shifting Cultivation Areas of NE States (WDPSCA) – 100 per centage

The flagship scheme in India for addressing shifting cultivation being implemented by State Agriculture Departments in NE has not been successful as can be seen from the Report of the Task Force on Strategy for Management of Shifting Cultivation 2003.

"The NE region is prone to a number of biophysical, institutional and socio-economical problems, resulting into subsistence agriculture with low input - low yield - low risk technology. The major problems confronting the agricultural growth of the region are enumerated below:

Acidic Soil: The acid soils of the region have low availability of phosphorus, which gets fixed on application, thus leading to low response to fertilizers. The soils have high concentration of iron and aluminum toxicity to plants.

High rainfall and humidity: The high rainfall and relative humidity not only creates favourable environment for wide range of pests, diseases and weeds but also create problems in their chemical control through spraying. Such a climate also creates problems in storage of grains and hay-making. The high rainfall and cloudy sky reduces the total sunshine hours so essential for food production.

Low temperature: Low temperature favours incidence of crop and animal diseases, resulting in low productivity. The low temperature prevailing for considerable period during winter limits the total growing period for crop production.

Undulating topography, hill slopes and altitude: The undulating topography, hill slopes and varying altitudes create problems of agricultural production. In many cases, varieties suited to low altitude areas do not perform well or are not suitable for medium and high altitudes. Consequently, separate sets of varieties of the same crop are needed for different altitudes. There are various technical and implementation constraints to agricultural development within the region, which vary according to agro-ecological situations.

Technical constraints: High level of diversity in the upland farming systems, in agro-climatic conditions (physiography, soils, altitude, rainfall, etc.) and in the socio-economic and socio-cultural conditions of the different tribal communities, precluding uniform prescriptions and requiring the development of location-specific solutions.

Limited opportunities for expanding or increasing the productivity of settled arable farming due to the limited availability of low lands suitable for wetland rice cultivation and the high cost of developing the remaining areas having potential for irrigation, declining fertility in permanent fields, limited availability of fertilizers and pesticides in the hilly areas, difficulty in promoting mechanized agriculture on sloping land, and low incidence of sunshine during the rainy season, resulting in low productivity of crops.

– Depletion of soil, water and forest resources in watersheds where short fallow cycles are practiced in shifting cultivation system.

- Relatively high mortality rate of domestic animals due to high incidence of diseases and insufficient animal health coverage.
- Limited availability or access to improved varieties of seeds/ planting materials and animals.
- High dependency of hill farmers on rice production as staple food causing difficulty in diversification of the cropping systems.

High population growth in the region results in increasing population density in the upland areas and more pressure on natural resources necessitating shorter fallow cycles, rapid encroachment on forest land and extending agriculture on steep sloping lands.

Implementation constraints: Lack of integration and co-ordination between different line departments of different states in the promotion of integrated development strategies and lack of experience of extension staff in participatory development.

Inadequate research, extension and training support services with little client oriented or participatory extension activities, resulting in communication gaps between extension agencies and the farmers.

- Lack of integration of research, extension and education in agricultural development programmes.
- Lack of adequate service facilities such as credit, input supply and marketing to support agricultural development activities.
- Dependency attitude of farmers who have continuously benefited from government sponsored subsidized schemes.
- Relative isolation and poor accessibility of most of the village settlements.
- Low level of understanding on the key aspects of indigenous farming systems including shifting cultivation without any scientific analysis of these farming systems.
- The transport and communication bottleneck and the resultant high transport costs in the region, limited scope for exploitation of development potential in areas like plantation and horticultural crops. The jhum areas are confined to areas where basic infrastructural facilities such as roads are not developed. In the absence of proper link roads, whatever produce are harvested

have to be brought to the 'main road' by head load which is not only uneconomical but lead to wastage of time and energy.

Environmental constraints

Ruggedness of terrain: The rugged terrain of the mountain ecosystem of NE India is not very congenial for development of settled cultivation because the technology and implements used in it are not ideally suitable to rugged terrain, slope and soil conditions.

Dry terraces: The dry slope terraces are generally rainfed and of smaller size in comparison to lower slope terraces which are usually irrigated and larger in size. Crop production is relatively lower in the former.

Micro-agroclimatic conditions of the flat valley lands: Soil moisture tends to be higher on the flat valley lands than in the hill slopes. It allows abundant growth of weeds and insect pests which are detrimental to the domesticated crops. Hence, there is problem of adaptation of high-yielding varieties (HYV), causing poor production.

Socio-economic and cultural constraints

Poor economic condition: Due to poor economic condition, the shifting cultivators, except a few, are unable to afford additional inputs like improved plough, fertilizer, pesticides, irrigation, etc. required for settled cultivation.

Lack of skill and work force: The practice of settled cultivation demands a greater skill, work load, attention and management from the cultivators, particularly from the male work force. But most of agricultural operations in hills are executed by women, causing shortage of skilled male work force.

Where do we go from here?

As is clear from the analyses made in the previous chapters, there is a distinct disconnect somewhere between the forest dwelling communities and the various government interventions, be it policy or programme. There is a necessity to bridge this gap with stakeholder participation in policy formulation and implementation followed by regular independent monitoring & evaluation.

The road ahead

There are several imperatives. Some of these are:-

- Demarcate and consolidate the Community Forests of NE India involving civil society. This should be a priority issue. Funds either from the Integrated Forest Protection Scheme (IFPS) or Twelfth Finance Commission can be mad available for the same.
- State officials (domicile to NE region) should go on Central Deputation to Delhi – the seat of power, particularly in those Central Ministries like Tribal Affairs, Department of Land Resources in Rural Development Ministry, Ministry of Agriculture, Ministry of Environment & Forests, and DONER Ministry. This will go a long way in fostering the spirit of national integration.
- Massive tree planting drive in degraded jhumlands. Although the centrally sponsored scheme on Watershed Development Project in Shifting Cultivation Areas of NE States (WDPSCA) has specifically provided that tree planting component will be given a preference while considering project proposal, the same is not given adequate attention by the States during implementation. In addition, the States must earmark some funds for focused tree planting programmes taking into consideration that for continued supply of timber to meet the ever increasing demand, natural forests alone is grossly inadequate now
- When the Central Government has specific programs for treating problem lands such as alkaline soils, why not a dedicated scheme for treating acidic soils, the major problem afflicting NE region. This will give a fillip to agriculture productivity, which is lowest in the country.
- As per recommendations of the National Forest Commission, within the broad parameters of the National Forest Policy, each State should have its own forest policy statement, for the sustainable management of its forest and wildlife resources. The Conference of Principal Chief Conservator of Forests and Chief Wildlife Wardens during 5-6 September 2006 has also reiterated that each State formulate its own Forest Policy highlighting their specificities. This is an important opportunity for addressing and recognizing the community forests of NE India. NGOs,

social activists, conservationists and the village community as stakeholders should be taken into confidence and duly involved.

- The 25 MPs (includes 1 from Sikkim) from the NE can play a very significant role in Parliament. They need to use their collective bargaining power by making a united stand on issues that are specific and of common concern to NE region. The North east MP Forum in Delhi should be strengthened.
- Bifurcate the Regional Centre of NAEB. Presently NEHU, Shillong serves as the sole RC catering to all the 7 States of NE which is inadequate since every State in the NE has its own University now. One more Regional Centre may be thought of – Manipur, Mizoram or Nagaland, taking into consideration the various parameters – law and order, connectivity, other logistics, etc.

References

Community Forestry International: Indigenous Forest Stewards of Northeast India – Mark Poffenberger, Editor, SK Barik, Dhrupad Choudhury, Vincent Darlong, Vishal Gupta, S Palit, Ivan Roy, Ibobi Singh, BK Tiwari, Sanjay Upadhyay.

Centre for Civil Society: Briefing Paper on Forest Policy Community Stewardship and Management *For the Terracotta Summit Beyond NEP 2004: Institutions, Incentives and Communities* March 7, 2005 - Somnath Bandyopadhyay, H B Soumya and Parth J Shah.

Eleventh Five Year Plan 2007-2012, Volume 1 Inclusive Growth, Planning Commission, Government of India.

Guidelines on Tribal-Forest Interface, Government of India, Ministry of Environment & Forests, New Delhi.

National Environment Policy 2006, Government of India, Ministry of Environment & Forests.

National Forest Policy Review, India by Institute of Forest Management.

Report of the National Forest Commission, Government of India, Ministry of Environment & Forests, 2006.

Report of the Task Force on Strategy for Management of Shifting Cultivation 2003, Government of India, Ministry of Environment & Forests.

6

Dwindling Forest Cover: Loss of Opportunities to Bioprospect Microbial Resources and Ecological Support Systems

S. R. Joshi

Man, with all his powers and prowess, depends on other biological entities for his normal function. This limitation has forced mankind to exploit the biological environment to his own advantage. This began with plants and animals and got extended to microbes and eventually to cultured animal, plant cells and tissues. Man had always benefited, either directly or indirectly, from and suffered due to microbes without ever realizing this fact. The history of microbes began with their first description by Anton Von Leewenhoek in 1677 and report of their involvement in lactic acid fermentation by Pasteur in 1857. Bread, cheese, vinegar, marinades, wine and beer have been made using fermentation by microbes and these microbes could belong to categories like yeasts, bacteria, molds and fungi. These processes are known for thousands of years. Commercial potential of microorganisms is immense since the scope of its activities covers the entire spectrum of human life and it draws on the capabilities of entire biological world and even beyond.

With regard to India and especially north-East India, research on microbiological resources has not been closely integrated with the problems and constraints confronting environment, forest resources, sustainable development and the low-income farmers in the agricultural sector. Biotechnology especially microbiology needs to be focused on some key problems within sustainable agriculture and rural development that historically have not been effectively addressed by conventional technologies. It has the potential in developing fast growing tree species, energy plantations and stress tolerant and cross-condition adaptable crops to benefit the population. Genetic engineering approaches have considerably broadened the range of gene pools which are now accessible for crop improvement, environmental monitoring, eco-friendly technologies and remediation purposes. If the region expects to benefit from modern biotechnologies, it has to give serious consideration to the drafting of bio-safety regulations that are tailored to meet the socio-economic needs. Bio-safety regulations and standards for risk assessment need to be harmonized within eco-regions since environments are common across political boundaries.

Man-Microbes-Forests Interface and resources demand of the day

With all the sophisticated technologies at his disposal, the modern man is currently in a state of euphorbia ignoring the fact that his livelihood is ultimately grounded in biological process enmeshed in intricate ecological webs whose destabilization could prove perilous to his own well-being. Industrialization and urbanization can be described as a conflict between human economic systems and the economy of nature. The former has been characterized by an apparently inevitable tendency towards concentration of capital and power through progress measured in terms of raw material extraction, production and trade. In contrast, the economy of nature is based on material cycling and capture of solar energy, which are highly, dispersed processes meeting the needs of millions of plant and animal species.

All species provide some kind of function to an ecosystem. They can capture and store, energy, produce organic material, decompose organic material, help to recycle water and nutrients. They can control erosion, control pests, fix atmospheric gases or help in regulation of climate. Under natural situation organisms live together influencing each other`s life directly or indirectly. Various vital processes such as growth, nutrition and reproduction depend very much upon the interactions between the individuals of the same species or between those of different species.

Pollination, seeds and fruits dispersal, grazing, parasitism, symbioses are the common examples of such interactions. Under natural conditions we find interdependence between animals themselves, between plants themselves, as well as between plants and animals. Moreover interdependence may also exist between species which are widely different, such as between trees and bacteria, or between cattle and birds, or between elephants and ants. The relationship between species may be beneficial to both, harmful to both or beneficial or harmful to one and neutral to the other.

India possesses a distinct identity, not only because of its geography, history and culture but also of the great diversity in the natural ecosystems. The varieties of forests harbour a rich biodiversity. The types of forests range from evergreen tropical rain forests in the Andaman and Nicobar Islands, the Western Ghats and the North eastern States to dry alpine Scrub vegetation in the Himalayas. Between these two extremes, the country has semi-green rainforests, deciduous monsoon forests, thorn forests, subtropical pine forests and temperate forests.

Forests environments have multiple uses and benefits of increasing value to our growing population. In addition to the well-known uses for wood, water, wildlife, range, and recreation, our forests enhance man`s general welfare. Among other benefits, forests environment provide shelter from climatic extremes, screens to hide unpleasant views, interesting and pleasing landscapes, and biological air conditioning. Environmental forestry offers service, production, and enjoyment for mankind.

Nature provided trees, the oceans, earth and the animals themselves as carbon sinks, or sponges. All organic life on this planet is carbon based and when plants and animals die, much of the carbon goes back into the ground where it has little impact on contributing to global warming Man`s industrial revolution that has now thrown the earth`s carbon dioxide sink out of balance. For example, our oceans have absorbed so much of carbon dioxide; they are becoming saturated and acidic. The problem is we have removed so many trees from this planet over the last couple of hundred years it`s going to take some time before the millions of trees planted in the couple of years mature enough to provide sequestration benefits.

Each plant has microbes associated with it and this may be in the form of associations like symbiotic, asymbiotic or even rhizospheric. The growth of plants is determined by the nature of plant growth promoting microbes. Proper soil management, agricultural practices, use of fertilizers which do not harm the microbes would in turn enhance better growth of plants and preserve microbial diversity.

Microorganisms are source of support systems derived by man in the form of the microbial processes and production. These include ethanol production, biogas, bioremediation, antibiotics to probiotics flora. Many more support that could be derived from microbial resources is yet to be explored as merely less than 5per cent of the microbial biota is known to humans till today. Destroying forest would mean annihilating the microorganisms associated with it and losing resources that are yet to be explored, documented and bioprospected.

Microbial wealth, Forests and North- east India

Although there are only few institutes who are involved in hardcore microbiology work in North-East India, however, the richness of biodiversity and the presence of wild forms and relatives can provide ample resources to exploit the microbial life forms and their engineering for:

- Biological control of pests;
- Protecting threatened species;
- Resurrecting extinct species;
- Cleaning up and managing the environment;
- Researching new products;
- Energy plantations for faster yield; and
- Mass cultivation and conservation of horticultural, floricultural and medicinal plants.

Genetic modification imparts new properties and capabilities in organisms, including plants, animals and microorganisms. There is an urgent need to protect sensitive areas like rain forests yet increase crop yields from the existing land. The rapid disappearance of species from the natural habitat due to human interventions makes it more important to look for alternatives to conserve them. One such alternative is biotechnology which enables to produce crops by reducing pesticides dependence, fertilizers dependence and irrigation and engineering to increase their resistance to plant diseases. The technology can be effectively utilized for toxin detection from water, air, food and soil to improve human health and soil productivity.

The emergence of bioprocess engineering may provide opportunity to explore plants compounds for use in detergents, nylon, glue, paints, lubricants, plastics, food colours. Again, the North eastern part of India being a hotspot of biodiversity offers enormous natural wealth of microbial life forms.

Agriculture, forestry, aquaculture and even mining of natural resources can be areas of interest for utilization and exploitation of microbiological developments. North east still holds its natural vigour and virginity in terms of biodiversity which can assist in increased crop yield , greater flexibility in growing environments, less use of chemical pesticides, improved nutritional content etc due to congenial environment. Biotechnology applied in many industrial products and processes brings with it significant environmental as well as economic benefits, in areas as diverse and important as climate change, water conservation and reduction in the production of toxic pollutants. Advances in areas such as genomics, proteomics, metabolomics and pathway engineering are delivering new generations of bio-transformation technologies opening up real potential for using renewable biomass as feed-stock for industrial products and processes across a broad range of economic sectors. Moving towards the use of more biobased products and processes offers the prospects of reducing reliance on finite resources of fossil fuels and developing an economy that is increasingly based on renewable resources.

Discussion

In North East India, Biotechnology research has not been closely integrated with the problems and constraints confronting environment, forest resources ,sustainable development and the low-income farmers in the agricultural sector .It needs to be focused on some key problems within sustainable agriculture and rural development that historically have not been effectively addressed by conventional technologies . The region has potential in developing fast growing tree species, varieties for energy plantations and stress tolerant plants and cross-condition adaptable crops to benefit the population. What is needed is Joint Forest Management, including farmers' groups in decision-making regarding the definition of sustainable agriculture and rural development objectives and defining proper forest policies. A dialogue between upstream public-sector biotechnology researchers and downstream on-farm researchers and farmers' groups are poor. There is an urgent need to device mechanisms for effective translation of farmers' expressed needs into research action through appropriate "participatory transfer". Furthermore, the situation is grim because there is no incentive mechanisms that would ensure that biotechnology research is targeted to the needs of poorer farmers or social groups.

The traditionally preserved forests based on religious rituals in North-East come alive with beneficial creatures that can naturally control pest and disease problems. In addition, the trillions of beneficial microbes work to adjust pH, make nutrients more available, improve plant health and increase crop quality and yields. These Forests (Traditional Forests like Sacred Groves for example) are a treasure definitely of the unexplored microbes of varied potential besides the wealth and knowledge of medicinal plants.

References

Bulock, JD (1987). Introduction to basic biotechnology. *In*: JD Bulock and B. Kristiansen (eds.) Basic Biotechnology, pp. 3-10. Academic Press. London.

Casida, Jr, LE (1966) Industrial Microbiology. Wiley Eastern Ltd. New Delhi (reprinted in 1991).

Joshi S.R. (2005), "Bioresource and Bioprospecting in North-East India: Unlocking the treasure for Socio- Economic development of the region", paper presented in National Conference cum workshop on Socio-economic development of India with special reference to North eastern India, 4-5 May 2005, Shillong College, Shillong.

Joshi, S.R. and Joshi, N. (2000). Man and His Environment. Gautam Bros. Meghalaya.

Kamboj V.P. (2004), "Traditional knowledge- patenting & benefits". paper presented in National workshop on conservation & sustainable utilization of medicinal & aromatic plants of North east India, 27-28 May 2004, NEHU, Shillong.

Lal, R and Lal, S. (1990). Crop improvement utilizing Biotechnology. CRC Press, Bocaraton, Florida(USA).

Natech, S., Chopra VL and Ramachandran S. (1987). Biotechnology in agriculture. Oxford & IBH, New Delhi.

Reddy K. Purshotham and Reddy D. Narasimha (2003).Environmental Education, Neelkamal Publications Pvt. Ltd., India.

Singh, BD (1998) . Biotechnology. Kalyani Publishers, India.

Street, HE (1977). Plant cell and Tissue culture. Blackwell. London.

Syiem D., Joshi S.R. and Syiem M.B. (2004), "Intellectual Property Rights and North east India: Issues and Relevance": paper presented In National Seminar on Intellectual Property Rights, 6-8 October 2004 ,St. Anthony's College, Shillong.

Willadsen, SM (1989). Cloning of sheep and cow embryos. Genome. 31: 956-962.

7

Man Forest Interface: The Scenario in North East India

Sanjeeb Kakoty

Whenever the question of man forest interface is put up, one picture invariably comes to my mind. The sleepy village of Semkhor in the North Cachar hill district of Assam. The year was probably 1993, and I was visiting my brother who was then working as a veterinary officer and posted at Maibong. With a borrowed bike we decided to visit the village of Semkhor, which literally translated means salt well. We made good time on the semi motorable road for about 2 hours, including a portion over an un-bridged river, when an avalanche over the road made us reluctantly abandon our bike and continue our journey on foot. It probably took us another two hours of trekking before we noticed a cluster of hamlets that seemed to squat in an irregular pattern. They dotted the entire face of the hill. This was the ancient village of Semkhor. Inhabited by the Semsas, a warrior clan of the Dimasa, the village seemed suspended in time. The people were semi naked and suspicious of outsiders. They lived off the forest, hunting, foraging and also engaging in jhum cultivation. They ate food out of leaves and occasionally bartered salt from their well for other necessities, with neighbouring tribes. According to oral tradition this village had been set up on a strategic location by the Dimasa Kings following their defeat in the war with the Ahoms and their retreat from Dimapur to Maibong in

1536 AD .The village was put in charge of the fierce Semsas and designed to act as a vanguard to check possible incursions into Maibong. It may be argued that if the village of Semkhor came into being around this time, it could boast of continuous human habitation for close to 500 years. In other words, for 500 long years the people have lived in complete harmony with the forest. Not as an adversary but as an integral part of the eco –system.

To this situation, welcome the British, and by the mere passage of a law these forests technically becomes reserved forests. In such a scenario, the inhabitants of Semkhor can overnight be rendered illegal settlers liable to persecution, eviction and jail sentences. Yes I am talking about the Forest Acts of 1865 and 1878.

With the British colonisation of India the resources of India including its minerals and forests were put at the exclusive use of the British Empire. This was sought to be achieved by monopolising trade and grant of absolute rights for the exploitation of resources. While doing so, it was not surprising that the indigenous population was legally barred from exercising rights over the forest and resources of their country! The beginning of the policy of reservation of forests and assertion of exclusive rights can be traced to the time when Malabar Teak was found to be the most durable material for making ships. By the beginning of the 19th century, the English Oak plantations used for ship building had been largely exhausted, thus prompting the search for alternative timber. The answer was Malabar Teak. Thus by 1806 an order was passed whereby the exploitation of Malabar Teak was reserved for the exclusive use of the Royal Navy. The bulk of the Teak was exported to England and at the same time dockyards were also set up in Surat and on the Malabar Coast.

Government efforts at increasing revenue collection ensure that huge swathes of pristine forests were cleared to make way for agricultural and commercial plantations such as tea. In time, the forests too were transformed into a source of revenue earning and commercial logging became a lucrative business proposition. In addition forests were also cleared for expansion of railway lines and roadways and what resulted was an unprecedented destruction of the Indian forests!

A few statistics, chosen at random, helps to form a more complete picture of the situation.

- 1860 – 1349 km of tracks had increased to 51,658 km in 1910

- In the Madras Presidency 250,000 sleepers or 35,000 trees required annually
- Betweenn 1869-1885, 6,500,000 deodar sleepers were supplied from the Yamuna forests alone
- Though only 6 species were suitable for sleepers 50 varieties were tried out.
- Each mile of railway construction required 860 sleepers. Annual requirement was around 1 million sleepers. Normal life of a sleeper is 12-14 years. Unsuitable wood lasted only about 3 years!
- In one instance, of 487 sleepers supplied 458 or 92 per cent was of unauthorised wood.

The imperial forest department came into being in 1864. For this purpose help was secured from Germany, which was supposed to be the leading European country in forest management. Dietrich Brandis, who was appointed the first Inspector General of Forests, had been a Botanist at the Bonn University. By 1865 the Indian Forest Act was passed. This was a hurriedly drafted piece of legislation mainly meant to facilitate government acquisition of forests required for railway supplies. This is considered to be the first attempt at asserting state monopoly in forest control. The Forest Act of 1865 enabled the British to acquire, demarcate and reserve forest area specifically for use in the railways. It also contained provisions under Section 8 which empowered the government to arrest without warrant anyone who encroached upon forest land demarcated for the purpose mentioned above.

A colonial mindset backed with requisite legislative provisions had enabled the British to carry out a systematic exploitation of the Indian forests. In a lot of instances, the rampant felling of trees could not be followed by the actual extraction of the felled trees. In such instances the timber, worth millions of rupees simply rotted away! As if this was not enough, new forest legislations with more stringent provisions were being contemplated. Promulgation of a new Act in 1878 enabled the government to exercise absolute control over tracts demarcated as valuable. In addition it was stated that the rights enjoyed by villagers over forests were not rights but privileges. It may be mentioned that rights are enjoyed by sovereign citizens while privileges were to be enjoyed at the mercy of the ruler! In other words, it was clear it was an assertion of British paramouncy in India in the light of which customary and traditional rights and privileges exercised by the individual and the community over forests, could only be enjoyed at their mercy.

Here it may be opportune to speak a word about the Sacred forests. Dietrich Brandis, the first Inspector General of forests had heaped praise on the extensive network of sacred forest in the sub-continent which he termed as the traditional form of forest preservation. He also praised the Indian rulers for their ecological awareness and in this he singled out the Rajasthan Chiefs who passionately preserved brushwood in a dry climate thereby setting a good example which the forest officials of the British government will do well to emulate.

Henry Skolimowski Professor of Humanities at the University of Michigan, who also holds the first Chair for Eco-philosophy at the University of Poland eloquently, describes the Khasi sacred Forests . He writes and I quote: "The Khasi do not have temples as we know them. For them the whole of nature is a temple; and they worship trees, groves, forests and rivers as deities, or as the places in which God resides. Because of their strong beliefs in the sanctity of nature, the highest peaks of the Khasi land have been saved from the axe." He goes on to refer the Imperial Gazetteer of India of the year 1908 which makes refers to the Khasi hills as being "clothed with clumps of oak, chestnut, magnolia, beech and other trees, which superstition has preserved from the axe of the wood cutter" He caustically remarks that What The Imperial Gazetteer calls "superstition" is now called Ecological Wisdom!

Prof Henry has made another interesting observation about the importance of sacred groves and forests for the people of ancient Greece and Rome, and how these sacred natural places were gradually eliminated with the advent of Christianity and secularism.

He notices a similar kind of predicament in the Khasi hills too, which saw a replay of the same problem with the creation of a scenario of the Khasi versus Christianity, or rather Christianity versus the traditional Khasi.

But there is nothing sacred about colonialism! Armed with stringent provisions of the Forest Acts, colonial India witnesses a massive destruction of forests be it clearance for construction of roads, laying of railway lines, building bridges and so on. As a matter of fact, India was a major source of timber supply during the World Wars. Between April 1917 to October 1918 228,076 tones of timber EXCLUDING railway sleepers was supplied by the specially created Timber Branch of the Munitions Board and 50,000 tones of fodder grass exported to help military operations in Egypt and Iraq. Between 1914 and 1919, 1.7 million cubic feet of timber mainly teak was exported annually.

In the meanwhile, the 1878 forest act was modified from time to time. Finally in 1927 a comprehensive Indian forest Act came into being. Under it, forests were classified into different categories each put under different levels of State control. The Act made it possible for the government to acquire any forest and classify it as a reserved or protected forest. This meant that all traditional rights over such land were automatically extinguished. In addition the Government could also prohibit animal grazing, mining and charcoal burning, stone quarrying and other similar activities. That such an act would rob traditional forest communities of their very means of livelihood did not seem to bother the government.

The irony of preventing Indians from using their own forests while the same were being plundered for use of Her Majesty's Government being the most pronounced during the World Wars. During the Second World War, in order to aid the British War Effort, a Timber Directorate was set up in Delhi in 1940. Government intentions were made amply clear and the Forest department was under instructions "to produce the maximum outrun possible", and it did not disappoint. By end of 1944, it had supplied 909,000 tones of timber. There seems to exist enough justification to believe that the basic purpose behind creating the Department of Forest as well as the Forest Research Institute was not to manage forests but to maximise timber extraction and to expand revenue from exploiting forest resources.

Post Independence Policy

The tenacity with which the Indian Legal System in the post independence era continues to cling on to rules, usages and laws that should have been a part of the country's colonial past is utterly astounding. The rules and acts that made up government policy and even the administrative structure for managing forests followed the blueprint laid by the British. For instance the Forest Policy Resolution of 1952 is almost a carbon copy of a similar document adopted in 1894. Not surprisingly the primary concerns of the policy is aimed at realizing maximum annual revenue and making available maximum forest resources for defence, railways and other such goals of national importance. The policy also continues the colonial legacy of asserting and re-asserting the primacy of the state as the ultimate custodian of all forests. This assertion of rights of the state is achieved by negating the rights of the community. The much awaited amendments to the Forest Act hardly delivered anything. For example, the forest (conservation) Act of 1980 and its Amendment of 1988 had the primary objective of transferring authority on forest utilization from the purview of State governments to the Ambit of the Union government. It was also made mandatory for the state to secure prior approval of the Union government before a reserved forest could be de-reserved.

Probably, what could have changed the course of forest management in India was the Scheduled Tribes and Other Traditional Forest Dwellers (Recognition of Forest Rights) Bill, 2006. For the first time in the history of Indian forests there was an admission of state culpability in denying forest dwellers their legitimate rights. It sought to go a step further to reverse this "historic injustice" and also give those communities an active role in forest management.

But the passage of the bill was anything but smooth. Strong lobbies espousing to represent the cause of Conservation, the Forest bureaucracy, business interests most notably of the paper and pulp industries raised numerous objections which included apprehensions that the law would distribute forest land to tribal families. Alarms were also raised that that wildlife and people can no longer co-exist, and all tigers would perish! The JPC recommended that the proposed 1980 cut-off year be changed to December 2005 and include all non-tribal traditional forest dwellers within its ambit. In addition to recognizing the rights of tribal and traditional forest dwellers in areas declared as protected areas; it also revised the process for identification of protected areas to ensure a more transparent process and increased the ceiling of 2.5 hectares on land to 4 hectares. Probably the most important recommendation was that no diversion of forest land would happen without the consent of the gramsabha or the village assembly.

Unfortunately for Indian democracy, the Government refused to place the JPC report in the parliament, citing serious differences on four major issues: cut-off date, inclusion of non-tribals, rights of gram sabhas and the ceiling issue. After much acrimony, a bill, ostensibly based on the recommendations of the JPC was finally placed in the Lok Sabha on 15th December 2006. A history of sorts was created when sixteen major amendments were moved by the tribal minister on the bill he himself introduced in the house! After heated debates and major amendments, the bill became an Act. It was a much mutilated cousin of the original draft! None the less, the noteworthy clauses included the right to homestead, cultivable and grazing land (occupied, and in use since December 2005), and to non-timber forest produce (partially, since the rights for the time being were limited to produces of 'plant origin' and fish). The Act also recognized, though in a restricted manner, the existence of legitimate non-tribal forest dwellers. Also given was the right to rehabilitation in case of past forcible displacement and the clause that all future notification of 'inviolate' conservation zones and curtailment of rights in Protected Areas shall require people's consent.. Perhaps the admission that the rights of forest dwellers included efforts at conservation of forests and biodiversity, and the call for people's involvement to strengthen conservation efforts are watersheds in forest management in India. Seen in conjunction with another provision of the Act which states that all forestlands-irrespective of location and category- traditionally used by communities would be henceforth treated as community forest resource, and forest dwellers can act decisively in conserving those resources, could lay the foundation for effective forest management.

But the real challenge before planners is to effectively co-opt the community as equal partners in forest management and conservation. Unless the commitment for conservation is total and absolute, the new policy cannot succeed. The skepticism of a section of the forest officials is understandable, as the only forests that have survived the onslaught of man have been mainly the reserved forests, protected with force and by excluding the community. But it must be admitted that the management of reserve forests too has not been without its share of scams and cover-ups. It is probably time to acknowledge the fact that comprehensive forest conservation can be achieved only by making the community responsible for it. Initial hand holding by the forest bureaucracy and learning to respect the wisdom of the community should help tide over the initial hiccups that are bound to occur.

Scientific Managment of Forests

With human population growing at a phenomenal rate, the tussle between forest conservation and the ever growing need for resource exploitation has reached breaking point. This fact was brought home to me a couple of years ago in Upper Assam when I along with my team were in the on a trip to film some of the last remaining rainforests of the region. Popularly called the Noadihing forests, it flanked the river Buridihing and included the Joypur reserve forests, and extended upto Digboi and

Arunachal Pradesh. During the course of our trekking deep inside the forests, it was perplexing not to encounter any animals. The reason soon became apparent when we came across teams of technicians, who claimed to be conducting surveys on behalf of the Premier Oil Company of Canada, who were setting off explosives deep in the heart of the forests, to ascertain the presence of petroleum. This had caused the animals to flee in panic. I was left hoping and praying that these surveys will reveal no oil, for if it did, it would be the end of the last of the rain forests!

To make matters worse, we also suffer from serious shortcomings in our so called scientific management of forests. We would do well to admit the fact tropical humid forest, is a complex web of life that may contain hundreds if not thousands of plant species, along with numerous species of micro-organisms as well as animals. Even today, we are yet to have a complete inventory of all the living organisms from a single hectare of any one tropical humid forest. Having an understanding of the relationship between them is light years away. Unfortunately, for a long time, the model of forest management adopted in India resembled the model adopted from the European experience of working in temperate forests.

A model drawn from the single species dominant temperature forests of Europe had no comparison to the bewildering species variety of the tropical forest. Who can be a better person than Dietrich Brandis to explain what they encountered. He pointed to the phenomenon of the low proportion of marketable species in any forest 'a peculiarity of forestry in India' The solution to this was a typical quick fix solution. Probably, no one in India can forget how the British, in order to prevent the weavers from Dhaka and Murshidabad, from competing with the mills of Lancashire, had the thumbs of the weavers chopped off! Similarly, with regards to the forest, they embarked on the journey of replacing the natural mixed species forests with forests consisting of only commercially valued trees. That such mono culture forests wrecked havoc on the eco-system was hardly of any concern.

With our typical Indian mentality of following the master we continue to pursue the same policies. As a classic example is the Bastar pine plantation project. In the 1970s a project was started whereby the tropical pine *Pinus carribbeana* would be planted in an area covering some 40,000 hectares in the tribal district of Bastar in Madhya Pradesh. Initially a 500 hectare plantation be developed in a span of five years on an experimental basis. At the end five years, work on the larger project started.

Undertaking of the huge project evoked massive opposition primarily from the effected tribals, who were staring at deprivation and poverty as the project would entail replacement of 40,000 hectares of natural forest from which they drew their sustenance. As is the usual practice, the central government appointed an expert group under the chairmanship of the inspector-general of forests to examine the project. When this committee requested data from the experimental 500 hectare forest, not a single data could be provided!

There is talk of making the North east the power house of India. Scores of mega hydel dams are proposed to be built in the next few years, which would produce hundreds of thousands of watts of power. What will be the ecological price for these dams? What would be the social costs? Would the small communities that inhabit these areas survive the onslaught of "development" to be unleashed by these projects? These are some of the questions that haunt many of us. That we were not alone in such thoughts was becoming increasingly apparent. The newspaper "The Telegraph" dated March 16 2009 reported that The All Idu Mishmi Cultural and Literary Society and the All Idu Mishmi Students Union had threatened to boycott the proposed public hearing for the 3,000MW Dibang Multipurpose Hydroelectric Project in Lower Dibang Valley district of Arunachal Pradesh. The report went on to quote some of the voices of the indigenous leaders who spoke on the matter. I quote it, as they have flagged the issue in its proper perspective:

"We want to see an end to the never-ending circus called public hearing for the project. We have the people's support and there will be a mass boycott of the public hearing as before. We submitted a memorandum to Prime Minister Manmohan Singh during his visit on January 31 last year, demanding immediate scrapping of the project which once comes up will push the 10,000 population of Idu Mishmi tribesmen living in the district to the brink. Singh laid the foundation stone of the mega project on that day in Itanagar, more than hundreds of kilometres away from the actual site fearing public outcry. Our associations will not budge an inch from our stand come what may," Mickrow said.

Echoing him, the general secretary of the Society, Mite Linggi, demanded that Lower Dibang Valley be declared as "ecologically-sensitive zone" according to the provisions under Section 3 of Environment Protection Act, 1986, claiming that the district had the highest forest cover in Arunachal Pradesh besides two wildlife sanctuaries and was part of the Dihang-Dibang Biosphere Reserve. He said Dibang Valley was also considered part of one of the 25 global biodiversity hotspots of the world nurturing rare and endangered flora and fauna and dense forests.

At least 17 mega hydro power projects have been proposed in Dibang Valley district to augment the hydro power potential of the country. "How can an influx of 6,000 outsiders as labourers and workers for a single project in a valley of fast diminishing 10,000 indigenous people be seen as cultural integration and assimilation? This is a cruel joke," Linggi said.

Probably, the concept of development itself, that stressed on growth rates and statistics rather than focusing on the human individual as the basic unit; that economic prosperity could by itself ensure true human development and happiness is certainly the most cruel joke modern man has imposed on humanity.

I. Barpujari H.K. (ed) The comprehensive History of Assam Vol II, Guwahati 1992.

II. Stebbing,E.P, The Forests of India, Vol. I p, 63, London 1922-27.

III. Albion, R.G, Forests and Sea Power, Harvard University Press, 1926.

IV. Gadgil, M and Guha,R, This Fissured Land, An Ecological History of India, Oxford, NewDelhi 1992.

V. ibid p 132.

VI. http://trumpeter.athabascau.ca/index.php/trumpet/article/view/357/561.

VII. Gadgil, et al, op.cit., Pp 138-140.

VIII. ibid., p 139.

List of Prposed Hydel Projects , "The Ecologist Asia" Vol 11 No.1 January –March 2003.

This is an indicative list of some major projects commissioned, under construction and proposed to come up in this region. Many of the proposed projects indicated here are in an advanced stage of planning and clearance.

Pakke (NEEPCO): 105 MW, proposed in the Preliminary CEA ranking. Preliminary survey being carried out. Reported to cause submergence within Pakke Wildlife Sanctuary.

Kolodyne (NEEPCO): 120 MW, proposed in the Preliminary CEA ranking.

Kaldan: 545 MW, proposed in the Preliminary CEA ranking.

Emra II: 870 MW, proposed in the Preliminary CEA ranking.

Etalin: 3,045 MW, proposed in the Preliminary CEA ranking.

Hutong: 950 MW, proposed in the Preliminary CEA ranking.

Kalai: 2,550 MW, proposed in the Preliminary CEA ranking.

Demwe: 3,000 MW, proposed in the Preliminary CEA ranking.

Noa-Dihing: 75 MW, proposed in the Preliminary CEA ranking.

Oju I&II: 4,505 MW combined, proposed in the Preliminary CEA ranking.

Kameng (NEEPCO): 600 MW (4x150), under construction. Two dams on the Bichom (96.5 m. high) and Tenga (60.5 m. high) rivers.The Bichom flow is to be diverted to the Tenga and further diverted to the Kimi powerhouse, bordering Pakke Sanctuary. The project was cleared in 2001 despite a grossly inadequate and inaccurate EIA report. Negative impacts already visible. Total land requirement = 710 ha. (all forestland). Directly affected population = 355 people (74 households).

Dikrong (NEEPCO): 100 MW, proposed. Under preliminary investigation.

Ranganadi (NEEPCO): 405 MW (3x135), already commissioned. 112 m. high dam on the Ranganadi will divert water into a reservoir on the Dikrong river (60 m. high dam) through an 8.5 km. long head race tunnel. Directly affected population = 600 people (Yazali village). Total submergence area = 1,640 ha. (unclassified state forest). No mention of impacts of inter-basin transfer on river regimes or consequent downstream impacts. Muck disposed into the Ranganadi during construction has severely affected the river and fish catch downstream.

Ranganadi II (NEEPCO): 130 MW (2x65), 117m. high dam. Submergence area = 988 ha. (80 ha. is cultivable land). To be located upstream of Ranganadi I since the firm power of stage I is only 65 MW. With stage II, this is likely to double. Feasibility studies conducted.

Upper Siang (NHPC): 11,000 MW, 257 m. high dam, proposed to be India's largest hydel project. Under survey and investigation, has received stage I site clearance from the MoEF, Total land required = 13,500 ha. of which 9,015 ha. is dense forest.

Middle Siang (NHPC): 700 MW, 154 m. high dam. Submergence area = 1,000 ha. Presently under survey and investigation and has received stage I site clearance.

Lower Siang (NHPC): 1,600 MW, 67 m. high dam. Total land required = 5,130 ha. (including 3,293 ha. of forestland and 123 ha. ofyftt/m/ orchard/unirrigated land). Presently under survey and investigation and has received stage I site clearance.

Upper Subansiri (NHPC): 2,500 MW, 265 m. high dam. Submergence area = 6,500 ha. Presently under survey and investigation and has received stage I site clearance.

Middle Subansiri (NHPC): 2,000 MW, 213 m. high dam. Submergence area = 6,000 ha. Presently under survey and investigation and has received stage I site clearance.

Lower Subansiri (NHPC): 2,000 MW, 116m. high dam. Submergence area = 3,436 ha (upto 70 km. upstream of the dam site!) of which 42 ha. is in the Tale Valley Wildlife Sanctuary. Presently undergoing environmental and forest clearance. Is likely to impact *beels,* dolphin habitats, wet rice cultivation and fish catch in downstream areas. The EIA has not reported these and several other impacts.

Debang (Brahmaputra Board): 3,000 MW. Is also meant to moderate the flow of the Brahmaputra in the Assam valley. Detailed project report was to have been submitted in December 2002. However, it is reported that infrastructural works including road construction have already started.

Rangit (NHPC): 60 MW (3x20), operational since 1999. 45 m. high, 100 m. long concrete gravity dam.

Teesta V (NHPC): 510 MW (3x170), 96.45 m. high, 182.5 m. long concrete gravity dam. Under construction. Submergence area = 326.66 ha. of land (122.17 ha. is forestland and 201.4 ha. is private land). 250 ha. of 'degraded forest' has been acquired for compensatory afforestation. 2per cent of total project cost kept for catchment area treatment. **Teesta High Dams** (NHPC): 332 MW, listed in CEA's preliminary ranking study.

Kaptai (Bangladesh Power Development Board): 230 MW, constructed in 1957,50 m. high dam, 600 m. long. Submergence area = 400 square kilometers. (including 16 square kilometers. of reserved forest, and 54,000 acres of arable land). Directly affected population = 18,000 tribal families, mostly Chakmas. There are reports of planned expansions to the project, which will lead to further negative impacts.

Pagladiya (Brahmaputra Board): 26 m. high, 23 km. long earth-filled dam. Granted environmental clearance. Submergence area=5,150 ha. in 26 villages. Progress on project slow due to opposition from local communities regarding loss of agricultural lands. Dam meant mainly for irrigation and flood moderation, will also generate 3 MW of power. Expected completion date: 2007.

Kopili (NEEPCO): 200 MW, already commissioned. Has caused the drying up of the feeder channels downstream that charge *beels,* the spawning grounds of fish.

Lower Kopili (NEEPCO): 150 MW (3x50), 71 m. high concrete gravity dam. Under construction. Located about 20 km. downstream of existing Kopili powerhouse.

Kopili Stage II (NEEPCO): Environmental and forest clearances not granted yet.

Gumti (state government): Presently generating 8.6 MW, commissioned in 1976. Submergence area = 46.34 square kilometers. of one of the state's most fertile valleys. Affected population = 8,000-10,000 tribal families.

Doyang (NEEPCO): 75 MW, commissioned. Has submerged 5,420 acres as per official records. Loss of lands has reduced the *jhum* cycle in adjoining lands to 4-5 years. 23 villages in the immediate catchment area are involved in a community-based catchment area treatment plan with the help of the central and state governments.

Tuirial (NEEPCO): 60 MW, 77 m. high dam. Target date of commissioning is June 2006. Granted environmental and forest clearance. Construction yet to begin. Submergence area = 5,380 ha, more than two lakh trees to be cut, 15 species listed in Schedule I of the *Wildlife (Protection) Act* found in the project area.

Tuivai (NEEPCO): 210 MW, 155 m. high dam. Submergence area = 1,600 ha. (55per cent is either open forest, grassland or permanent *jhum* and the remaining is dense vegetation). Environmental clearance granted. 1,60,000 mature trees to be felled in the submergence area alone, to be 'compensated' for by the creation of Lengteng Sanctuary and bringing old growth *jhum* area under the forest department.

Tipaimukh (NEEPCO): 1,500 MW (6x250), 162.8 m. high, 390 m. long earthen core, rock-filled dam. Meant to also moderate floods. Forest and environmental clearance not yet granted. Faces stiff opposition from indigenous peoples who stand to lose their traditional homelands.

Loktak (NHPC): 105 MW (3x35), 10.7 m. high and 58.8 m. long barrage. Operational since 1983. Natural water flow disrupted impacting the health of the lake and the *khangpok* people. Agricultural lands submerged and fish catch reduced.The sangai deer and other fauna in and around Keibul Lamjao National Park are threatened. The lake is a wetland of global importance under the Ramsar Convention and an Important Bird Area. Loktak Project Affected Areas Action Committee still seeking adequate compensation.

Loktak downstream (NHPC): 90 MW (3x30), 64.5 m. high, 250 m. long earthen dam. Aims to utilise tail-race discharge of existing Loktak project and free draining catchment of the Leimatak river. Environmental clearance granted but construction yet to begin due to security concerns. Submergence area = 257 ha. including 170 ha. of wet paddy.

Bairabi (state government): 80 MW, 62 m. high, 180 m. long earth-filled dam. Could provide 180 km. waterway from Aizawl to Silchar. Land required = 9,294 ha. (7,721 ha. is forestland and the rest is valuable horticultural land).

Myntdu-Leishka (state government): 84 MW (2x42), 59 m. high dam. Granted environmental clearance, forest clearance pending. Land required= 181.16 ha. (102 ha. of forestland). Will drown forests inhabited by tiger, jungle cat and binturong. Mining continues upstream threatening the lifespan of the dam due to acidic water.

"The Telegraph" March 16, 2009

8

Towards a Green Society-Alternative Framework for a Socio-Environmental Movement

Jyotirmay Goswami

Economists must redefine poverty as a shortage of biomass rather than a shortage of cash. Gross natural product is more relevant to the poor than gross national product.

Anil Agarwal, environmentalist.

Natural wealth has counted as an indispensable asset in almost all cultures through centuries. In the last few centuries however, wealth of a different kind has gained prominence. Instead of relying on gross natural products, nation-states have been singularly occupied with the gross national product. The effects of this single-minded pursuit are there for all to see. The matters of widespread environmental degradation, extinction of biological species at an unprecedented rate, climate change and global warming have now become common knowledge. We have come to know that the rapid strides of apparent human progress made since the industrial revolution have come at a terrible cost. In India, the upbeat spirit of the green revolution has given way to the sobering realization that indiscriminate use of chemical fertilizers, pesticides, groundwater and special varieties of seeds create a dangerous imbalance in the natural world which, if the present trend continues, would become uninhabitable for future generations. At the same time, our quest for efficiency through automation has led to centralized control of resources, inequality and disruption of livelihoods of a large section of the population. People all over the world are becoming surplus everyday. Life in the villages is becoming less and less viable due to intrusion of distant economic forces. This has resulted in steady migration of people from villages to towns and from towns to cities, where one has to fall in line with the prevailing culture. Just as agriculture has been taken over by a single, resource intensive mode, the space for diverse cultures is set to be taken over by a single, resource intensive culture promoted by advertising and mass media. Expulsion of man from his natural and cultural context has meant deprivation from his sources of spiritual sustenance which cannot be compensated by material possessions. Human relations have come to be polluted by the compulsion to judge everything against an impersonal, utilitarian standard. When the scale of this disarray is considered together with the disarray in the natural world, it becomes amply clear that the path currently traversed by mankind is not sustainable. One has to look for a better way of life.

Return to organic ways of agriculture appears to be inevitable. However, the transformations of the last few decades have meant that such a switch would result in a swift decline in productivity and income in the short term. Even though the long-term outlook is better, the prospect of short-term loss prevents people from taking to organic farming. The transition can be made possible by diverse farming. By resorting to various supplementary plants and crops, it is possible to augment the income a bit. However, the issue of security can be overwhelming in the short term. Any departure from conventional wisdom would attract suspicious attention. Any possibility of success of an experimental path would generate neighbours' envy.

These hurdles are natural consequences of a piecemeal approach which has been our bane. We would not have been in the present mess if we had realized how interconnected human life is with the lives of other humans and the natural world. Every effort to get out of this mess has to be part of a holistic approach. The present article is about a proposal to reclaim our lost greenery in a way that might strengthen village life and complement organic and diverse farming.

Greenery in India – history and present status

Forests in India have traditionally been common property. Local people were de facto owners of the forests. The economic benefits reaped by them were far more than that accruing to the local king. The present status of forests as exclusive state property goes back to the British era. As a matter of fact, much of today's forest policy that has resulted in timber-intensive, fruit-free reserved forests is the legacy of the pre-independence period. A comprehensive Memorandum of the Government of India brought out by the then Governor Lord Dalhousie on 3rd August 1855 – a document which has since been dubbed as the Charter of Indian Forestry – clearly proclaimed forests as state property, so that the public cannot come in the way of logging for railway sleepers. The *Indian Forestry Act* (1865) gave the government full control over felling of trees in forests. The first *National Forest Policy* adopted in 1894 sought to demarcate and reserve forests in the interest of the state.[1] The National Forest Policy of independent India declared in 1952 turned out to be only a variant of the 1894 policy. Clause 7 of this policy detailing 'claims of neighbouring communities', denied forest dwellers the right to gather forest produce, in the name of national needs (which included 'sustained supply of timbre and other forest produce required for defence, communications and industry' and 'realization of maximum annual revenue in perpetuity'). Since then, the local people have consistently been perceived as a nuisance as far as forests are concerned. This perception continued even when resource utilization came to be supplemented by forest conservation in *sarkari* policy. It is clear that the attitude of the local dwellers was seen by the Commission as a cause of the problem rather than as a natural effect of government policy. Such confusion is bound to occur when analysis is made without a sense of history. Before the government turned to wholesale exploitation of forest resources, it was the local people who had protected the forests. Local communities intended to remain where they were for centuries. They had to ensure sustained liveability of their surrounding area. Conservation of forests *had to* be imbibed in their way of life. The tribal peoples' tradition of festivals, their yearly ritual of tree-planting – all these were aspects of an organic relation they had with the forest. Government policy has put increasing pressure on this relation over the years.

Despite the lack of sensitivity in government policy, there had been some individual initiatives to involve local people in the affair of forests. In 1971, Dr. Ajit Bandyopadhyay started a project on Socio-Economic

Forestry in the village Arabari in Medinipur district of West Bengal. In this model, local people were engaged in guarding government-owned forest in exchange for economic benefits in the form of limited use of forest products. The economic incentive benefited those who were engaged in this manner, while the limited objective of protecting the forest was also met. This model came to be followed in several states, with notable success in West Bengal and Madhya Pradesh.

In any case, 'success' of experiments like Arabari had an impact on the second National Forest Policy of India (1988), which called for joint management of forests. It was the first time in independent India that the role and interest of dwellers of the forest and adjoining areas were incorporated in official policy. The new policy mentioned three types of forests: reserved forest, protected forest and village forest. The third category consists of small and scattered plots of community owned forest. In spite of extensive planning and expenditure for these three categories of forests, the amount of greenery remains inadequate. According to the Forest Survey of India, greenery accounted for 19.49per cent of all land in India, and 9per cent in West Bengal in 1991. By 1997 the per centage in India fell to 19.27 and that in West Bengal increased to 9.41. The figure given by the government of West Bengal is 19.98per cent – bolstered perhaps by inclusion of bushes, sugarcane plantations and anything else that looks green. In any case, these per centages are much less than the estimated 40per cent a century ago[7] or even the desirable level of 33per cent envisaged in the National Forest Policy 1988. Village forests account for most of the loss of greenery. There have been some official effort to reclaim greenery outside reserved or protected areas, with the aim of increasing the supply of timber and firewood so that there is less pressure on protected and reserved forests and some revenue to the panchayats. These schemes for 'social forestry' have succeeded in meeting these objectives to some extent. However, production of timber and firewood can hardly be called *forestry*. 'Social forestry' has never been a *social* phenomenon either. The local panchayat and the handful of guards are benefited financially, but the rest of the society has absolutely no role in the matter.

A government initiative would always have its limitations because of its scale and mode of operation. If need-based involvement of a few local individuals in a government initiative have brought about positive changes, organic involvement of the society at large in local initiatives holds the potential for much more.

Sustainability in agriculture – the road traversed

Agriculture has been the basis of Indian society for thousands of years. Naturally the tools and techniques used here have evolved over a long period of time. David Hopper describes the agricultural practices in an Indian village – Senapore – in the mid-1950s: "The age-old techniques have been refined and sharpened by countless years of experience, and each generation seem to have had its experimenters who added a bit here and changed a practice there and thus improved the community lore. ... In general each man comes close to doing the best he can with his knowledge and cultural background."

The agricultural methods employed in those days are not just a reflection of traditional wisdom. For centuries, both the size of a community and its agricultural practices had to be in steady equilibrium with natural limits. The expansion of communication and large scale economic activities since the nineteenth century led to migration of an ever-increasing portion of the population out of their native places. The population began to grow, along with explosion of its needs. Instead of checking the unsustainable rate of population growth, the five year plans brought about policies that would enhance migration on one hand and rapidly increase food production on the other.

The Green Revolution was born. Heavy tractors dug out soil from hitherto unreachable depths. Pesticides and fertilizers were used in liberal measures. Groundwater flew out of pumps where irrigation canals could not reach. High yielding varieties of seeds were introduced. Crop yield increased steadily. India became 'self-sufficient' in food grains.

The ill-effects of the revolutionary agricultural methods have begun to surface after a few decades. Production cost has risen steadily while crop price has reduced in real terms in an increasingly global market. There has been erosion and loss of fertility of soil. The groundwater level has sunk to record depths in various parts of the country. Many tricks of the trade that optimally utilized local resources have also gone extinct.

Outline of the greening initiative

In summary, the present situation is paradoxical in many ways. Government schemes for social forestry can neither include nor benefit the society as a whole. Greenery is vital for healthy physique and graceful spirit of the society, but apparently there is not much land left for this to happen. The idea, in a nutshell, is that the whole society would participate in an intensive greening effort from which everyone would be benefited directly. The fact is that even today there is plenty of unutilized land here and there where one can grow fruit trees, vegetables and other plants of economic value.

Autonomous greenery

Like all plants and animals, man has all along shared land with others. If someone locks up a piece of land for his exclusive use, he not only deprives others but also himself of the fruit of that ownership. The land may belong to the government or to a private owner, but the fact of that ownership need not prevent its greenery from being shared or socially overseen. Just as there are multiple layers of territories used by different life forms in a forest, farmland can also be used for multiple purposes. As far as the greening initiative is concerned, it is the society that would have central role in the pooling of resources, production and distribution – within the existing pattern of ownership and obligation. Government funding or intervention would not be necessary. In this domain, the community would have autonomy. The greenery would be autonomous.

Autonomous greenery would also be a personal matter in a special sense. Recognized (if limited) ownership of a tree or a plant can give a person context and scope to give shape to his or her idea in a creative way. The owner having a personal bond with the trees can enjoy autonomy in a small, private domain.

Initial hurdles – how to overcome them

The greening initiative signifies a collective confidence reposed on the viability of our villages. It signifies a mental switch from a passive to an active role. There are many who would like to use this occasion as a turning point and make another switch that they would have been contemplating for some time – moving from 'high-tech' farming to organic farming. It is now well established that organic farming gives more return in the long run than 'high-tech' farming. However, if one turns overnight to organic farming then there would be a temporary fall in the return, mainly because of the damage done to the soil by years of using pesticides and chemical lizers. This is indeed a difficult hurdle.

n obstacle to diverse farming as well as the greening initiative, is e to cattle. The prospect of the wandering cattle straying into a field ng hopes can be enough to prevent people from starting anything need for additional vigilance can also be seen as opportunity for community work. There is another serious difficulty in growing diverse plants and crops. Over the years, a great deal of traditional knowledge – on how to tend to these different forms of greenery simultaneously – has almost passed into oblivion. This knowledge is invaluable because it is specific to local geography, climate and culture. One may have to turn to wiser heads to regain part of that knowledge as well as the confidence to manage numerous subtle aspects of farming.

By far the most difficult obstacle to the greening initiative is the existence of all kinds of conflicting interests. The tale of pursuance of self-interest – even at the expense of others – is as old as human history, and this is unlikely to change. As natural conditions for habitation get steadily disrupted by large scale economic activities, as plant and animal species become extinct with a speed that surpasses even that of the extinction of dinosaurs, as the human souls are caught by droves out of their social and cultural habitats as if by invisible fishing nets and dumped in a uniformly mechanical environment, the isolation of man becomes increasingly complete. In this context, Collective will – and the urge to touch every soul and carry everyone along – can win over the situation.

Individual opposition is not a serious problem, but consorted opposition is. When a whole community of people is guided by a selfish motive, it is rather difficult to win over the entire group. Sectarian feelings may have existed in the past, but it was never as widespread and long lasting as it is now. The vast majority that stands to be left out of the process forever is faced with a unique dilemma. They are powerless if they are isolated, but coming together under the banner of the party would at best bring some of them a bit of material relief – they would not get back the shade of the tree that they have lost.

Greenery and harmony – some precedents

The nature of the proposed greening initiative is different from other government or non-government initiatives in many ways. Yet the idea has resemblance with experiments undertaken elsewhere and practices followed at another period of history. In pre-British India, the forest was deemed as common property. The common man was as much responsible for its protection as was the zamindar or local king. This notion continues in some parts of India even today. Society still protects *Deovani* forests in Maharashtra. The same can be said about *Oran* forests of Rajasthan. Living trees are just not felled in these areas. The commitment of the *Bishnoi* people of Rajasthan towards protecting forests and wildlife is legendary. The people of the *Mahar* caste in pre-British Maharashtra protected forests and ensured judicious use of forest products. Generations of *Mahar*s have played this traditional role assigned to them by society. In West Bengal, tribal people specially protected forests designated as *Zaheera Than* or forests of God. Such forests are gradually shrinking into oblivion. Trees – especially fruit trees – in localities and fields are found in abundance in various pockets of north and south India. Greenery has a special place in society in Kerala. Diverse kinds of trees – including those for fruits, spices and flower – can be seen in between agricultural plots as well as residential areas. One can build on these experiences and traditions.

Experiments with socially initiated greenery have taken place outside India too. The example of Cuba shows how social participation can materialize in the face of a crisis and bring in remarkable change in the well-being of the common man. Because of stiff trade sanctions imposed by USA and its allies, Cuba's economy depended heavily on trade with the soviet block countries. The example of Cuba is a special one. The exceptional social unity observed there was precipitated by an immediate sense of crisis shared by everyone. The government had also played a crucial part in mobilizing the people, who are mostly urban. These circumstances are quite different from the present Indian situation, although increasing hardship on the part of the general populace may eventually lead to an acute crisis. However, the Cuban experience shows us that human resources can once again reign supreme over material resources. This example can still serve as a beacon for the greening initiative. Harmonious union of various sections of society lies at the heart of the greening initiative proposed here. The legacy of harmony is very much a part of Indian culture.

Potential gains

A major advantage of the greening initiative would be that, by supplementing income and food and by providing a conducive atmosphere, it would give the requisite impetus to organic and diverse farming. There would be renewed hope of arresting the steady degradation and erosion of soil brought about by indiscriminate use of chemical fertilizers and pesticides. Production of chemical fertilizers and pesticides requires oil and other ingredients apart from energy. Lesser use of these chemicals would mean reduced depletion of finite resources and less pollution from industrial production. Pollution would also be reduced directly by the planting of trees.

One of the ill-effects of indiscriminate cultivation of rice has been a steady and unsustainable depletion of groundwater. The number of shallow tube wells in West Bengal alone increased from 20,000 in 1980 to 600,000 in 2004; the number increased to 10,00,000 by 2006. Several thousand litres of water are needed for producing one kilogram of rice. Lifting of groundwater requires energy, and much of the irrigating water does not trickle back to the underground aquifer. Turning to diverse farming of carefully chosen crops would reduce the demand for water. On the other hand, organic farming means abundance of organic matter in the soil, which has greater capacity to retain water. Trees planted in the vicinity would also bind the soil and help retain water. Some additional ponds may be dug to hold rainwater to ensure additional supply of water for growing trees and crop, where necessary. Diverse farming would go hand-in-hand with the greening initiative. In fact, planting of trees *is* a form of farming. The greening initiative signifies this change of mind-set, which is a necessary precondition for stability of returns on investment.

Symbiotic raising of crops, plants, animals and fish would involve concerted participation of people at all levels. This would arrest the widespread trend of people becoming surplus in their natural habitat, generate employment throughout the year, and make the conditions in villages more liveable. Thus, a reverse migration from towns and cities to villages would be promoted. Renewed faith in the village and renewed focus on farming is vital for a green society. There would be renewed opportunity for crafts and small scale industries that centre around agriculture. Folk arts and handicrafts that draw sustenance from a close tie with nature can spring forth in thousands of forms. If this happens, there would be lesser need for government schemes to patronize or preserve these art forms as relics of a lost civilization.

The greening initiative and the associated cultural shift would promote self sufficiency and would strengthen the local economy. New opportunities for local investment, including small scale agro-industries, micro-banking, schools and health centres would be created. A major advantage of investment in the local community is that the return does not always have to be immediate or even monetary. If the investment leads to stability of the community and enhanced goodwill for the investor, who along with his loved ones stands to be benefited well into the future, these gains count as returns on investment.

For the community, autonomous greenery would be an attempt to reclaim some of the ground lost to forces beyond its control. It can adjust the scale and pace of change to suit the capacity of individual members. Through confluence of intertwined rights, shared natural wealth and collective opportunity, people can attain a sense of contentment which is perhaps the goal that lies behind the various ideals of equality.

Engagement of everyone in the society is a crucial aspect of the greening initiative. Abundance of fruit trees and vegetable plants would ensure a steady supply of fruits and vegetables of different kind round the year. Many of these items can be eaten directly or with minimal cooking. Apart from savings on fuel or firewood, there would be positive effect on health due to increased presence of vitamins and minerals in the diet. Another indirect health benefit may come about if fruit snacks can recover some of the space captured by fast food.

In countries like India where the climate is hot almost round the year, working under the blazing sun is not easy. In the past, there used to be some fruit and shade trees even in the agricultural fields. These trees have disappeared with increased pressure on land. Intense pressure of the global market has meant that every available bit of land has to be utilized for the prevalent crop of the time. As a result, the small farmer or wage earner has to toil away in scorching heat, sometimes risking their health and even life. Many proponents of human rights have little idea as to exactly how harsh this workplace is in comparison to air-conditioned auditoriums and cars. A workplace with adequate shades should count as one of the fundamental rights of every human being. Those who have had their 'right to shade' violated can regain it through controlled expansion of greenery. How welcome for them would be a little break for meal under the shade of a small tree in the field itself! The shade of a tree is a treasure.

A section of the academic world can also benefit from participating in the greening initiative. Exploring easier and more effective ways of expanding greenery can be the subject of research. Seed banks. Comparative studies of various practices suitable for different conditions can be undertaken. If the study touches the lives of real communities, the academicians behind it would surely have a rare sense of reward. Given the huge diversity of land, climate and other conditions in different parts of the country, there is considerable scope for such studies.

Glimpses of a green society

In many ways man has lost his links with the natural world. A green society would strive to recover this loss. The society could celebrate the re-established ties with nature through a festival. Many indigenous societies already have such festivals.

The villagers in the green society would not have to look upon the distant town for their livelihood. Livelihood itself would be less dependent on formal employment. The crop in the field, the fish in the pond and the plants, trees and animals raised in every available nook and corner would demand a great deal of attention and affection from the people, who in turn would have a source of healthy sustenance. The wealth of a green society would consist of life in numerous forms. This wealth would lie everywhere – in the fields, ponds, rivers and in the little hutments. It would be seen in the fingers of a villager adding another stroke of paint to the picture of the green society. It would be seen in the intense expression of the villagers' collective will. This wealth cannot be secured by individual or hired effort, as it does not fit nicely into bank lockers or to the confines of a walled compound. Security can only come from close ties of the society and of human relation. The green society is his only hope for escaping the lethal cloud of comprehensive pollution that looks set to consume the world around him.

Welcome are all earth's lands, each for its kind,
Welcome are lands of pine and oak,
Welcome are lands of lemon and fig,
Welcome are lands of gold,
Welcome are lands of wheat and maize, welcome those of the grape,
Welcome are lands of sugar and rice,
Welcome cotton-lands, welcome those of the white
potato and sweet potato,
Welcome are mountains, flats, sands, forests, prairies,
Welcome the rich borders of river, table-lands, openings,
Welcome the measureless grazing-lands, welcome the
teeming soil of orchards, flats, honey, hemp;
Welcome just as much the other more hard-faced lands,
Lands rich as lands of gold or wheat and fruit lands,
Lands of mines, lands of the manly rugged ores,
Lands of coal, copper, lead, tin, zinc,
Lands of iron – lands of the make of the axe.

Walt Whitman.

References

Barton, Gregory (2000). Keepers of the jungle: Environmental management in British India, 1855-1900. *The Historian*, Vol. 62, No. 558 (March 22), pp.557-574.

Bhattacharya, Prasanta (1986) *Burdwan University Science Journal*, Vol. 3, No. 1, p.88.

Dabholkar, Shripad A. (1998). *Plenty for All*. Mehta Publishing House, Kolhapur

Douthwaite, Richard (1996). *Short Circuit*, Lilliput Press.

Dreeze, Jean and Sen, Amartya (1995). *India: Economic Development and Social Opportunity*, Oxford University Press.

Goswami, Paramesh and Sengupta, Debasis (2006). *Pragati Marichika*, Visva-Bharati.

Government of India (1988). *National Forest Policy 1988*, New Delhi, Ministry of Environment and Forests. http://www.envfor.nic.in/nfap/detailed-policy.html

Government of India (1952). *National Forest Policy for India*, New Delhi, Ministry of Food and Agriculture Resolution. Document downloaded on April 11, 2006 from http://forest.ap.nic.in/forestper cent20policies/1952per cent20Forestper cent20Policy.html

Guha, Ramachandra (1983), Forestry in British and post-British India: A historical analysis , *Economic and Political Weekly*, Vol. 18, Issue 45, 29th October (Part I) and Issue 46, 5th November (Part II).

Mishra, Tapan (1998). *Banglar Bon Jangal.* Pashchim Banga Bigyan Mancha.

New reports, *Bartamaan*, 16th and 24th June, 2004.

Schultz, Theodore (1964). *Transforming Traditional Agriculture*, Yale University Press, New Haven, CT, USA.

Seabrook, Jeremy (2003). A short history of the future. *The Statesman*, 20th April.

Sinclair, Minor (2001). From big to small, toxic to green: new strategies to grow food in Cuba, *Revista*, Spring 2001 (Issue on Food in the Americas), http://www.fas.harvard.edu/~drclas/publications/revista/food/Sinclair.html

9

Land and Forests Rights of Tribals in West Khasi Hills of Meghalaya

Gideon L. Kharkongor

West Khasi Hills is one of the districts of Meghalaya situated west of the capital city of Shillong. It has an area of 5247 Squere kilometers and a population of 296049 persons with a population density of 56 peoples per Squere kilometers (2001). In other words, West Khasi Hills District has 23 per cent of Meghalaya's area and 13 per cent of the state's total population. The district has the distinction of the district with the highest proportion of scheduled tribe population in the country. The region is rich in its diversity with forest cover of 2080 Squere kilometers i.e. 40 per cent of the total district area and 22 per cent of the state's total forest cover. The livelihood of the people is mainly agriculture while a substantial proportion of them are engaged in trading in forests products. In recent times, the people have exploited the forest for timber, fire wood, medicinal plants, orchids and other forest products.

The forest area in the region is largely under the category of unclassed forest and where the ownership is private or with the clan or with the community. The Reserved forest in the state, as a whole, is only 10 per cent of the total forest area and just about 4 per cent of the geographical area. West Khasi Hills has only a negligible area under the forest department. The typical land holding system that exists in the state and the region has only limited the control of the government in the overall development of forest. Administratively, there is a forest department of the government and the Autonomous District Council that is empowered to look into the affairs of forest in the state. Apparently, the Khasi Hills Autonomous District Council (KHADC) plays a dominant role in the affairs of forest in the state. Since forest is privately owned, the task of issuing licenses for the trade in timber and forest products lies with the KHADC. The KHADC collects royalty and this by itself becomes the reliable source of information on the volume and quality of trade of forest products from the region.

People and forest interface

The chief occupation of the people in the region is agriculture where 50 per cent of the population is engage in agriculture and its allied activities. The other activities includes mining, forestry, fishing which are classified in the 'other workers' category while there is no major industry in the region. A small per centage of the people are engage in household industry like weaving, black smithy etc. Shifting Cultivation or 'jhum' is the farming method adopted in the region which is a traditional method of land rotation rather than crops. The forest is cleared by cutting and burning before it is ready for cropping. There are a large number of families (jhumias) dependent on jhum. The age old method was sustainable till the time when land and time was available in plenty to the people. In the last few years, due to the fragmentation of land and the growing pressure on land, the jhum cycle was drastically shortened and this has led to the degradation of land especially the forest land.

The region is rich in biodiversity where different species of flora and fauna is found in the region. The tree species found here are the sal (shorea robusta), pine, teak, birch, schim, beach, nahar, champs, bamboo, titaopa, gamari, makrisal and others. There are a variety of orchids, medicinal plants and other forest produce that include thatch grasses, sun grasses, roofing leaves, dhoop wood, wild pepper, chalmukra, solonum berries, charita, socunmatara, ronfalfia, syrpentina, tejpata, ritha, neem and others.

The forest products are classified under major and minor forest produce. Timber is a major forest produce while grasses, medicinal plants, herbs, *tezpata*, broom etc are classified as minor forest produce. Charcoal is included in the category of minor forest produce. Timber is traded to feed the numerous sawmills and wood based industries in the state. The government has issued licenses to these industries and many more are running without valid licenses. Truck load of timber is seen taken out of the forest. The District Council issues permission for private owners to felled trees by a nominal payment of loyalty fees. The Supreme Court ban on felling and trading of timber in 1996, based on the civil writ petition filed by T. N. Godavarman Thirumulpad against the Union of India has resulted in the decreased of felling trees. It has also resulted in a large number of people being unemployed. Some scholars have estimated that the per centage of people turning unemployed directly or indirectly comes to about 70per cent. The forest owners and traders persuaded the government for an alternate solution, in such; the government of Meghalaya submitted an affidavit that a large number of people are forced to unemployment due to the enforcement of the ban. The Supreme Court in 1997 has, however, clarified that charcoal does not come under the purview of the ban since it is a minor product. This has resulted in an even perilous situation since people have the option and an alternative in the production and trade of charcoal.

Besides these, the people trade in medicinal plants and herbs which fetch a handsome price in the markets within the state, outside and to Bangladesh. Mining of few minerals such as limestone, coal and peat is taking place while uranium is yet to be extracted. Broomsticks and bay leaves (tezpata) are supplied in abundance from the region. Cattle and goats rearing is taking place within the forest area. During winter, when fodder is not available, the cattles are kept in the forest for they will feed on their own.

Table: 1: Land Use Classification in WKH, 2004-05

Category	(in Ha)	per cent
Area Under Forest	207960	40.2
Land not available for Cultivation	70437	13.6
Other un-cultivated land excluding Fallow Land	152097	29.4
Fallow Land	65974	12.8
Net Area Sown	20632	4.0
Total	**517100**	**100**

Source: Statistical Abstract Meghalaya

Figure 1: Land Use in West Khasi Hills

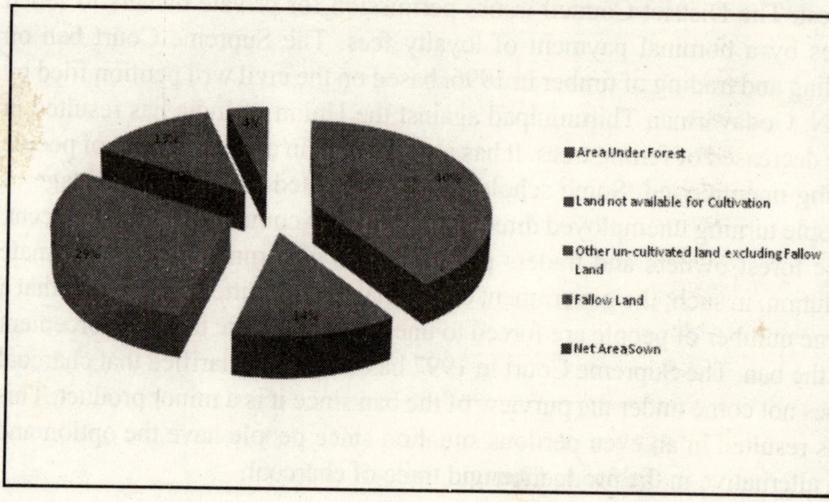

Table: 2-Forest Area by class in Meghalaya

Forest by Class	Total area in Sq.km	Per centage to total forest area	Per centage to total geog. area	Ownership status
Reserve forest(Govt forest, National Parks & Sanctuaries)	993	10.46	4.43	Government
Private forest	384	4.04	1.71	Syiem, Lyngdoh, Nokmas
Protected forest	179	1.88	0.80	Government, District Council
Village forest	25.9	.28	.12	Community
Community (Raid) forest	768	8.09	3.42	Community
Unclassed	7146.5	75.25	31.86	District council
Total forest	**9496.4**	**100**	**42.34**	

Source: Chief Conservator of forest, Meghalaya.

Table: 3- Forest type distribution of Meghalaya

Forest type	Forest Area	per cent to total forest land
Tropical Moist Deciduous	4100.75	49.68
Sal	2095.19	25.39
Sub-tropical pine	1010.52	12.24
Tropical Semi-evergreen	392.87	4.76
Sub-tropical evergreen	362.52	4.39
Tropical evergreen	219.77	3.54
Total Forest Area	**8253.62**	**100**

Source: S.Sharma

Table: 4: Changing Forest in Meghalaya

Change in (per cent)
87–89 89–91 91–93 93–95 95–97 97–99 99–01 91–01
–2.52 0.73 –0.33 –0.17 –0.18 –0.08 –0.16 –0.39
***Source:* Ministry of Environment and Forests, various years.**

Land and Forest interface

The tradition-oriented rights of the tribal people of India on lands and forests are not only the indigenous issue developed long ago, but in most of the cases these surpass the frontiers of history. After their many-sided confrontations with the contemporary conceptions, the question of traditional rights and their feasibility has raised a lot of queries and counter-queries in the sphere of safeguarding the indiscriminate destruction of the forested environment and unusual utilization of the lands in nature. The question has not only been confined to the academic circle, but it has stirred conspicuously the political domain of the country very recently. The tribal people living in and around the forests are not the destructor but the conservers of forests and related natural resources. Their love and sympathy for the natural resources have been processed characteristically in their cultural patterns and ultimately these have oriented the all-embracing tribal philosophy.

Forest has been an important source of livelihood for the people of the West Khasi hills of Meghalaya providing them not only fuel, fodder and timber but also food, fruits and medicines. For thousands of years the Khasis have lived on gathering both timber and non-timber products from the forests. With sharp insight gained through close and sustained interactions with nature, they have also learnt to adapt new modes of production to their forest life. Jhum, that is slash and burn or shifting cultivation, and horticultural plantations are two examples of the adaptation where the tribes efficiently use the basic skills of agriculture to grow food crops and fruit plants in their forest lands.

The Scheduled Tribes and Other Traditional forest Dwellers (Recognition of Forest rights) Act, 2008 of the Government of India to assigned ownership of minor fcrest produce of the people living in and around the forest will be highly scrutinized. The fact of the matter, still remains, that people in the region are dependent from the forest produce and only sustainable resource utilization coupled with community responsibility and government policies will help to ensure that the people have a harmonic life with forest. Since the unusual and unique land system that exist in the region, the people will have to be given the larger responsible and be the stakeholders in the various schemes and policies of the government. Indeed laws exist and government or the district council will have to enforce them, but involving the community should be the utmost priority for any soluble solutions. The private and community lands should be given emphasis for tree growth and the same can be converted to preserve and protected forest land like the sacred grove, which has been a traditional biodiversity practices of the people within the state and the region.

Environmental Impact

Industrial use

About 43 per cent of wood produced is consumed for industrial purposes alone. The average utilization of wood from Indian forest is estimated at 50 per cent and thus if 22 million m3 is extracted the total cut from forest could be double. This would mean that not all felled trees are utilized but are left unused in the forests. The forest department have issued licenses to 41 sawmills which include 6 in West Khasi Hills. Though this is an official figure, but one who visits the district will see at least two dozens of sawmills operating in the district. Since the ban on timber the royalty collected is reduced and permission to trade on timber was on old stock only. Very often with the pretext of clearing the old stock, new felled trees are also traded and is evident from the truck-load of timber still being transported to different areas coming from the region.

However, it was acknowledged that charcoal is a minor forest product. Supreme Court in its order in 1997 clarified that there is no ban on charcoal production. Herein lays the crux of the problem. Taking advantage of this loophole unchecked deforestation is taking place for charcoal production. The charcoal produced in Meghalaya is an important industrial fuel apart from being used as domestic fuel for cooking and heating purposes. In the industrial sector large amounts of charcoal is used in foundries and forges, in the extraction and refining of metals, especially iron, and in numerous other metallurgical industries. Private forest owners who are all indulged in charcoal making do not replant forests to the same level as they were first found. After a clear-cut, the only trees planted are ones that will produce lots of wood. This replanting is better coined as planting of a tree farm, which does not support the same animal and herbal plants that forests had, leading to a loss in biodiversity.

In West Khasi Hills, approximately 1 lakh people are employed directly or indirectly in the production of charcoal. According to District Council report an equivalent of 24,915 tons of charcoal / wood was consumed in 2003-2004 with the value of Rs.99.66 million. Around 63 MT of charcoal is transported from only one area of West Khasi Hills every week to Shillong and Byrnihat, the latter where the ferro alloy industries are located. Ferro Silicon alloy units use charcoal as the main raw material for production along with quartzite. Easy availability of both quartzite and charcoal in Meghalaya influenced the setting up of ferro alloy industrial units in Byrnihat. Ferro-alloy is made in a submerged arc furnace where carbon in the form of coke or charcoal is used as a reducing agent to obtain a specific grade of ferro-alloy. To produce one ton of ferro Silicon 1.4 to 1.7 ton of charcoal is required. In Meghalaya the ferro alloy units produce close to 700 to 1000 ton of ferro Silicon alloy per day. Electricity availability also plays a vital role in the operation of these industrial units. Power availability for 13 to 14 hours a day allows an output of 500 to 600 tons of processed ferro alloy. According to the available production figures to be 600 ton / day, it is estimated that the total amount of charcoal used in this industry alone amounts to 960 ton / day. The industrial purchase of one ton of charcoal is Rs. 3700 to Rs.4000, implying the total cost of 960 ton of charcoal to be Rs. 40, 00,000 .Production of one metric ton of charcoal requires about 8m3 of wood (8m3 x 63 = 504 m3) .This indicates that deforestation of the forest in the region is proceeding at an shocking rate of 50.4 ha/week or 2621 ha/yr.

Meghalaya Charcoal (Control of production, storage, trade and transit) Rules, 2008 came into force with effect from 2008 with allowance for disposal of existing stock. The use of charcoal in these industries has dropped since then but charcoal from the region and other regions of the state are still feeding the ferro alloy industries. There are alternative to wood charcoal as bamboo charcoal are said to be more viable both economically and environmentally. The working scheme of the government has taken care of these grave issues and one hopes that with such scheme a ray of optimism will instill in the people. Firewood use has not been put under any of the clauses and this too has encourages people to go for commercial trade in firewood. This has become a grey area for both policy makers and forest traders who incidentally have not given much notice.

Sustainable Forest management

Since the UNCED in 1992, a lot of emphasis is laid on the sustainable use of forest resources with the responsibility on the managers of forest on the impact on forest through excessive utilization of forest resources. Research, monitoring, and modeling can play vital role in providing information on biodiversity management which is part of sustainable management of natural forest.

Joint Forest Management (JFM) is set up to shoulder responsibility to the fringe forest users of their roles for afforestation of degraded community forest. JFM started in Meghalaya in 2003 where there are 73 JFM committees managing 4000 ha of forest area. The National Afforestation and Economic Board have approved Rs 12 crores under TFDA for the implementation of National Afforestation programme in Meghalaya. West Khasi Hills Social Forestry Division under the Forest Development Agency in Meghalaya (FDAM) has approved Rs. 143 lakhs in 2005 for the regeneration of forest in the district. The paradigm shift in forest policy from conservation, production and commercial oriented to the global concern now for environmental stability, maintenance of ecological balance and atmospheric equilibrium has warranted enactment of new laws concerning forest which threatens the relationship of tribals with their land.

West Khasi district have experienced the intervention of IFAD livelihood projects in transforming *jhumias* into other sustainable livelihoods. This success story of IFAD can be replicated to forest land where people will be able to sustain livelihood besides preserving the forest and forest products.

Validation of medicinal plants and herb is the need of the hour since local medicine men depend on the forest for their medicines and herbs. Local practitioner should be protected as vested business party has exploited the rich biodiversity of the state's forest. Forest being in the concurrent list, the state forest department can only be a watch dog with little or no powers of protecting the forests of the state.

Discussion

Forest plays an important role in the world today. They are important environmentally, economically, socially and ethically. It is the provider of food, medicines, water, clean air and others. We need to conserve and preserve the forests. Tribals in West Khasi Hills and the state as a whole have realized the drastic change of climate and the availability of forest resources for their livelihood. If forest will sustain them any further, a pragmatic and holistic approach is needed both from the people and the government. The government and the district council will not only be concern of its royalty collection and its revenue, but need to adhere to stringent laws that already govern the management and protection of forest in the state. The Meghalaya Forest (removal of Timber) Regulation Act, 1981 , Meghalaya Forest Regulation Act,1973, Meghalaya Charcoal (Control of production, storage, trade and transit) Rules, 2008 are few laws within the state to curb the excessive deforestation but to allow limited usage of forest produce to the local indigenous people for their livelihood. The Schedule Tribe and Other Tribal Forest Dwellers (Recognition of Forest Fights) Act, 2006 will empower the people in the district to have more powers over the forest products which actually is already envisaged in the National Forest Policy 1988. The situation in the state of Meghalaya is already grim with forest rights belonging to the community; such laws will only deteriorate the already degraded forest of the state.

The tribals should be taken into confidence in forest management programmes and any attempt of its conservation will largely depend on them. Urbanization has grown rapidly at the expense of both agricultural and forest land. Forest land is largely concentrated in rural areas where the largest number of tribals inhabits. Sustaining any more forest area will solely depend on the nature of life the tribals live. The best strategy for the government and the district council is to include private, clan and community land for programmes like JFM and Eco-Reserve Areas so that more and more land can be brought under forest. Government has, hitherto, little role in management and growth of forest since the forest under government is negligible. Hence their energy and resources can be diverted to the community forest for its growth and development. This can be summarized by what our Father of the Nation said, 'after all nature has enough for everyone's need but not to everyone's greed'.

Acknowledgement

My heartfelt gratitude to the Chief and Assistant Chief Conservator of Forest, Government of Meghalaya, the Deputy Chief Forest Officer, KHADC, Project Manager, NERCOMP-IFAD, Nongstoin and P.Madhavan, CSE,New Delhi for all their invaluable inputs.

References

Agnihotri, S. K., Dutta Ray(2005). Perspective of Security and Development in North east India, Concept Publication, New Delhi.

Dallmeier & Comiskey (1998). Forest biodiversity research,monitoring and modelling: Conceptual background and old world case studies, Man and Biosphere Series,Vol.20,Parthenon Publishing, Carnforth.

Dutta Ray, B. (2000): Population, Poverty and Environment in North east India, Concept Publication, New Delhi.

Dutta Ray,B. (2002): Agenda for North east India, Concept Publication, New Delhi.

Dutta Ray,B.and Alam K.(2002): Forest Reserve of North east India, Omsons Publishing, New Delhi.

Dutta,B.B.(1987): Land Relations in North east India, People's Publishing House, New Delhi.

Earl, (1976) in Robert, Burgess (1978): Journal of Forestry, vol.76.No.3.Society of American Foresters.

Kumar,S.(2002): The Medicinal Plants of North east India, Scientific Publishers India, Jodhpur.

Nongbri, Tiplut (2001): Timber Ban in North-East India, Effects on Livelihood and Gender. Economic and Political Weekly, Special Articles, May 26, 2001.

Rai,R.K.(Ed)(2001): Environment Resources and Development. GSNEHR (India) NEHU, Shillong.

Rao N.Subba (2001), Sustainable Forest Management: Some reflections. Kurushetra, Vol.49, No.6 March 2001.

Sarkar, R.M. (2006): Land and Forest Rights of the Tribals Today. New Delhi.

Sarma, S. (2003): Meghalaya: Land and Forest: A Remote Sensing Based Study, Concept Publication, New Delhi.

State of Forest Report, 2005, Forest Survey of India.

United Khasi-Jaintia Hills Autonomous District (Management and Control of Forests) Rules, 1960.

10

Crisis in Man Forest Relation: A Serious Concern

Sumana Paul

Man is an integral part of physical and biological world in which he lives. In a broader sense, his physical and biological world may be defined as environment which consists of biotic (living) and abiotic (non-living) components. Nature had evolved a wonderful pattern in forest environment. Forest is not only a collection of trees, herbs and shrubs, but it is also a complicated structure which is a result of combined influence of climate, soil, biotic interference, topography and time. A slight change in any of the factors may alter and damage the structure of the ecosystem. Forest environment provides ecological balance and forest remains as the rich natural resource of the mother earth for mankind. Forests are the most vital component of the biosphere and the source of basic resources as well. That is why it is called GREEN GOL.

Ever since the arrival of man on the surface of the earth, he has been either living in the forest or surviving on the forest. Hence man and natural environment have cordial relationship from time immemorial. Our Country has a very rich past of Vedic culture which was born and developed in the forests. Saints of ancient times had realized the importance of various plants that were useful for the survival of man. So the farsighted Rishis included such plants in religious worship. It is evident from available records that they have done it only to preserve the plants. Even today the tradition of tree worship is prevalent among different communities in our country. The Rigveda advocates regular plantation and it was treated as a sacred religious duty at that time. Society at that time had been very much aware of the fact that indiscriminate destruction of the plants and forests would result in disease and pollution of atmosphere. It had also been realized during ancient time that like plants and trees, animals and wildlife are also very useful for the survival of man. So they protected the animals too.

Slowly and gradually man devised the knowledge of science and technology and improved his standard of living. But at the same time he started destroying and exploiting nature in the name of progress and prosperity. There is nothing wrong in optimum utilization of our natural resources, but unfortunately that has not been the case as has been experienced over a long passage of time. Infact, man started cutting the forest and exploited the nature for satisfaction of his greed rather than meeting his basic needs resulting into environmental degradation. The poorer section of our society mostly depends on forests for food, fuel and other requirements, thus apparently destroying the forests resources to a large extent.

The North East scenario

The north east (NE) region bestowed with abundant natural resources is no doubt the richest region of the country in terms of forests cover and bio-diversity value. The region comprises 8per cent of the total geographical area of the country. But entire north east region consists of 25per cent of the country's forest cover makes it one of the richest bio-diversity zones of the country.

According to the latest findings of Aaranyak (an Assam based NGO, on 6th June 2008) the large scale loss of forest cover in the NE in recent times is due to deforestation caused by human encroachment on forest land. The findings also show that 64per cent of the total geographical area of the NE (2, 55,000 sq.km.) is currently under forest cover. Two decades earlier it was 84per cent.

The forests contain several species of timber which have high commercial value. Besides timber a number of non-timber forest products like bamboo, cane, mushrooms, orchids, important grass species, honey, wax and important medicinal plants are found in the forests. Meghalaya is only state in the north east India where Pine trees grow luxuriantly in her hills. Pine wood provides the people with the timber for house building, for furniture making, with fuel wood for cooking and with charcoal for keeping the living rooms and working places warm, particularly during the long winter months. The forests of the NE are enriched with wonderful variety of the wildlife which includes wild elephants, buffaloes, deer, snakes, different varieties of colorful birds, butterflies and world famous one horned rhinos.

The main issues of environmental crisis of the region

1. **Population, Poverty and Illiteracy**: Population growth is a significant factor of environmental degradation. The entire region is highly populated. As per 2001 census NE accounts for 3.7per cent of the country's population. Population growth becomes a burden as it multiplies pressure on space and resource. Unchecked population growth has aggravated the problems of poverty and illiteracy. Data shows that only in urban areas approximately 30per cent of the population lives Below Poverty Line (BPL). The corresponding figure in rural areas is much higher. Statistics also reveals that the percentage of illiteracy among the rural female of north east is really alarming. Unless sincere efforts to increase the percentage of female literacy is made one cannot control population. These factors directly or indirectly contribute to environmental crisis.

   Again agriculture and its allied activities have extremely grown in order to fulfill the hunger of the population. Excessive use of chemical fertilizers and pesticides in extensive cultivation to increase the output has led to soil degradation and other environmental problems.

2. **Industrialization and urbanization**: Industrialization is another significant cause of environmental pollution. The fertilizers industries, refineries and paper industries of north east part of India discharge sewage and industrial effluents which contain organic pollution and other chemicals. This results in air pollution and water contamination.

   Urbanization is also causing environmental problems. Urbanization creates infrastructural pressure and population concentration. Due to growing urbanization the size and the number of cities/towns increases as a result of which competition for production and consumption and goods and services and space intensifies. Urbanization should be planned otherwise it would be counterproductive. The growing urbanization in north east India is not only due to natural increase of population in urban areas but also due to migration of people from rural areas and from smaller towns to the cities.

Again, larger the size of the population, greater will be the volume of waste created. Domestic waste production is another main problem. Improper disposal of this waste results in diseases like diarrhea, malaria, etc.

3. **Deforestation and Environmental Pollution**: Forests are essential for environmental protection. Deforestation causes many environmental problems. It causes floods, droughts and so on. The droughts lead to problem of water scarcity, wind erosion and the like. Because of cutting forests in hills and plains draught has become a perennial feature in India. Even Cheerapunjee, which receives the maximum rainfall in the country, has a problem of water scarcity, because forests have been destroyed (Ushai Rai, 2000).

   Besides, growth of population, urbanization and forest fire, jhoom cultivation is again another important cause of deforestation in hill states of north east India. Jhuming is a wasteful and laborious method of agriculture. Moreover, it is an ecologically harmful practice. It leaves the land barren and forest area shrinking. The cumulative effect of such human activities practiced since past has been devastating on environment. It renders the soil bare and easy target of erosion during heavy monsoon rains. The top fertile soil which took thousands of years to form is washed away with few showers and deposited in the river basin. The rivers in the plains thereby become shallow causing frequent flood.

4. **Plastic Bags and Environmental Pollution**: Use of plastic especially plastic bags, pollutes the environment. Plastic bags are non-biodegradable. So when they are thrown away they create many environmental problems. Streams and rivers are affected by plastic bags. Some plastic bags contain toxic dyes. When the plastic bag is used for packing the food stuffs, the toxic dyes can leak into food stuffs endangering consumer health (Bharati Chaturvedi, 1999).

   Plastic cannot be continuously recycled. Since most carry bags are made of recycled materials, its further use to the reprocessing industry is limited. Burning results in air pollution and toxic ash. Now the central and the state governments have taken steps to check the use of plastic bags, but unless "WE" participate actively in stopping the use of plastic bags, nothing will happen.

5. **Pollution and Health**: The harmful gases discharged by the vehicles, as emission burning of firewood and disposal of garbage causes serious health hazards. In cities, the major source of pollution is transportation and its brown-black fumes. Due to all these factors it is difficult to get fresh air for breathing. There is acute shortage of drinking water and whatever little water is available is severely contaminated by pollutants.

   Unscientific mining is also another cause of environmental degradation, soil erosion and health hazards. The government cannot do anything against the persons found engaged in such type of activities. Public interference is required to deal with such kind of activities.

6. **Man Elephant conflict**: Another important environmental crisis in north eastern region is man-elephant conflict. Assam and Meghalaya have a projected elephant population of 7000 (reported in India Times, 13th November, 2008). Out of 7000, about 2000 elephants stray around without home. Undated destruction of forests has posed serious consequences to man and wild animals and elephants have suffered the most as they need larger areas for survival. Destruction of forests for construction of roads, buildings and railway track have compelled elephants to venture out to human habitation in search of food causing tragic deaths to both man and elephant.

Available figures present a frightening depiction of wide spreading man-elephant conflict. In Assam, over 600 hundred people have been killed by wild elephants in the past 16 years at an average of 40 a year. Wild elephants also destroy a large number of paddy fields and demolish houses. A few years ago 4 wild elephants were poisoned to death by harassed villagers in Sonitpur district of Assam. Man-elephant conflict in Assam and other north eastern states are burning problems for the villagers as they have to spend sleepless nights during the harvesting seasons. So appropriate steps should be taken by the concerned authorities to conserve the wild elephants. The NGOs and the nature lovers should come forward and awareness campaign should be launched among the common people to protect the elephants of north east region by restoring their habitat and provide them adequate space with enough food.

Educational intervention

Crisis in Man-Forest relation is the greatest crisis of the day. Human activities today on the earth led to rapid deterioration of the environment and science and technology will not be able to prevent degradation of the environment. Educational intervention through Environmental Education programmers in formal and nonformal setting can be one of the important remedy.

There has been a lot of discussion regarding the tremendous value crisis in our society that is being faced today. Looking into the value crisis it is an imperative need to make the education of our younger generation value oriented. Environmental awareness is also a kind of value and should become a part of everyone's value system. So it is necessary that a person should live in the atmosphere of environment consciousness. It is indeed a difficult task to make people environment conscious only through teaching the subject, so it is necessary to create an atmosphere of environment consciousness. Once we are successful in making ourselves and our younger generation environment conscious then probably we may be in a position to save and protect our environment.

The chief objective of environmental education (EE) is that individual and social groups should acquire awareness and knowledge, develop attitudes, skills and abilities and participate in solving reallife environment problems. EE is an integral process which deals with man's inter-relationships with his nature and man-made surroundings. EE is intended to promote among citizens the awareness and understanding of environment, our relation to it and concern and responsible action necessary to assure our survival and to improve the quality of life.

EE should be integrated in the curricula at all levels. Childhood is the most formative period of development and at this stage importance should be given on developing love for nature. Small children may be given lesson in saving the environment during prayer time. They should be given the impression from the very beginning of their life that God gave them responsibility and authority to care for their earth. Teachers should plan some class room activities. Small children should be taught things like –

- Use public transport
- Switch off the lights

- Avoid poly bags
- Not to waste water
- Not to spit on road
- Not to throw garbage here and there.

In the Secondary level and further higher level along with teaching, students should be involved in various environment related activities, since it is an accepted fact that students learn better when they are personally involved in activities related to the subject. Moreover, difficult concepts relating to EE can be explained and given a practical touch by taking the students out of class room as a regular practice on field trips, excursions, visit to a zoo, eco-development camps.

An analysis of the important causes of environmental crisis in the region reveals that rapid growth of population to a large extent is responsible for the crisis in man forest relation in north east region. Hence, population education needs to given top priority along with EE for checking the environmental crisis. Population education is not only about population awareness but also about values and attitude towards environment. To save our environment we need to educate our rural masses, especially the women folk about the dangerous consequences of high population on our earth and how to save our earth either by organizing village meetings or through adult non-formal education.

Some suggestions for minimising crisis

The following suggestions are given to overcome the crisis of man forest relationships in north east India.

- In order to control population growth the age of marriage for girls should be raised from 18 years to 20 years and for boys 25 years from present 21 years. Measures should be taken to increase the female literacy rate especially of the rural female.
- The expansion of employment opportunities and establishment of some recreation centers in rural areas will bring about a reduction in migration rate from rural to urban areas. This will reduce the urbanization and also greatly solve the problems of water scarcity, sanitation problems in urban areas.

- The government and voluntary organizations should take steps to popularize the use of bio-fertilizers and bio-pesticides which are environment friendly. Steps should also be taken by the government for controlling air pollution due to emission from automobiles. There should be some enforcing agencies run by the nature lovers to see the implementation of various laws enacted by the Governments.
- Voluntary organizations should come forward to celebrate "World Environment Day" not only in urban areas but in the rural north-east to create awareness of rural masses about the environmental crisis and encourage them in environmental conservation activities. But celebrating Environment Day once in 365 days and not doing anything in another 364 days will not serve the purpose. Regular awareness campaigns should be conducted by voluntary agencies by nature lovers to create awareness among the general public.
- Various agencies of mass media like Radio, Television Channels and Cable Operators should come forward to present serials stressing of the need for environmental protection. Environmental messages should regularly appear on many products right from bottled water to paper bags to make people environment conscious.
- The NSS and NCC units of various schools and colleges can be given responsibility of spreading the news of environmental protection among the rural folk and creating public awareness about relevant loss and policies among general public through street dramas and poster campaigns, distribution of colorful pamphlets and stickers etc.
- Seminars, workshops should be organized at regular intervals not only for students, educationists and scientist but also for common people to discuss environment related issues.
- The idea of community gardening should also be popularized among the rural folk. Villagers should also be involved in tree plantation programme and they should be helped to identify the possible area of tree plantation (like roads, riversides or barren lands).

- At this point of time it is very important to revive our primitive culture. Reflections of such cultural values need to ensure through text-books, stories, poems etc. If we really want to save our nature we must take lessons from the past.

The crisis in the environment needs to be addressed immediately. This crisis to a large extent is a man-made phenomenon. So to prevent the crisis PEOPLES PARTICIPATION (PP) is essential. If we really want an improvement in man-forest relationship we need to educate our younger generation towards developing a positive attitude towards our natural heritage.

References

Bharti Chaturvedi, (1999). Plastic Bags, Hindu Survey of the Environment.

Chhokar, K. B. (2000). Sharing Concerns in Environmental Education. University News,38(26):1-3.

Environmental Information System, Assam Science Technology and Environment Council, www.envisassam.org

Meghalaya Biodiversity Conservation and Dams, Ministry of Environment and Forests, Press Releases, June 26, 2009.

Nanda S. K., (2000), Indian Education and Its problems Today, Kalyani Publishers,New Delhi.

Newsletter, Soil and Water Conservation Meghalaya, Vol. 6, No. 6, July – December 2003.

Ushai Rai. (2000). Eating into the Green Reserves,Yojana,August, P.45

11

Man Forest Interface: Sharing of Experiences

M. C. Malakar

Why we want forests? What are the services provided by forests? These are some of the questions which may stir one's mind before he or she convinces himself or herself to protect and conserve the forests. In brief we may categorize the services provided by the forests in four categories as below:

1. Economic Services (Material Production): Forests produce timber, fuel, fodder, fruits, building materials, medicinal plants, raw materials for many products like the processed wood, pulp & paper, match boxes, rubber etc. Forests also provide the grazing ground for the village cattle.
2. Ecological Services: Forests produce Oxygen, without which the animal kingdom, including the human being, could not have survived or existed. Forests sequester Carbon from the atmosphere, regulate the water regime, maintain the soil quality, reduce soil erosion, maintain the atmospheric temperature and modulate the climate. Forests are also the storehouse of genepool.
3. Sociocultural Services: Many people have strong cultural and spiritual attachments to the forests, without which it would be difficult for them to think about their survival.

4. Scenic and landscape Services and values: Forests also have aesthetic values because of its scenic and landscape beauty.

It is, therefore, the ecological services of the forests, basically for which we need to protect and conserve our forests for our own survival. However, a common man is primarily interested in the materials produced by the forests, which lead to increased biotic pressure, making it unable to bear and resulting in degradation and denudation of the forests.

While the proportion of land to be kept under forests would obviously vary in different areas depending upon local/regional/climatic conditions, keeping in view the various factors, the National Forest Policy envisages that the country as a whole should aim at maintaining one-third of its total land area under forests, though the hills and the mountainous regions should aim to maintain two-third of the areas under forest cover for preventing erosion and land degradation.

Under this background, the forest cover in the north east, as revealed from the 2005 Forest Survey of India Report (produced below), does not appear to be very grim or alarming.

State/UT	Geographic Area	Very Dense Forest	Moderately Dense Forest	Open Forest	Total Forest	Per cent of G.A.
Arunachal Pradesh	83,743	14,411	37,977	15,389	67,777	80.93
Assam	78,438	1,444	11,387	14,814	27,645	35.24
Manipur	22,327	923	5,541	10,622	17,086	76.53
Meghalaya	22,429	338	6,808	9,842	16,988	75.74
Mizoram	21,081	133	6,173	12,378	18,684	88.63
Nagaland	16,579	236	5,602	7,881	13,719	82.75
Tripura	10,486	61	4,969	3,125	8,155	77.77
Total	**255,083**	**17,546**	**78,457**	**74,051**	**170,054**	**66.67**

However, in view of the fact that government hardly exercises any control over the forests other than the Reserved Forests and the Protected

Areas, the scenario relating to the legal status of the forests, as revealed from the following table, does not appear to be very encouraging.

State	Geographic Area	RF		PF		UF	
		Area	per cent to GA	Area	per cent to GA	Area	per cent to GA
Arunachal Pradesh	83,743	10,546	12.59	9,528	11.38	31,466	37.57
Assam	78,438	17,864	22.77	-	-	8,968	11.43
Manipur	22,327	1,467	6.57	4,171	18.68	11,780	52.76
Meghalaya	22,429	1,112	4.96	12	0.05	8,372	37.33
Mizoram	21,081	7,909	37.52	3,568	16.92	5,240	24.86
Nagaland	16,579	86	0.52	508	3.06	8,628	52.04
Tripura	10,486	4,175	39.81	2	0.02	2,117	20.19
Total	**255,083**	**43,159**	**16.92**	**17,789**	**6.97**	**76,571**	**30.02**

Furthermore, comparison of the forest covers of the north east region during 2003 and 2005 (*Forest Survey of India Report, 2005*), as depicted in the following matrix, indicates the necessity of taking strong measures to avert the negative trend.

Forest Cover Change Matrix for the North east Region (in square kilometers)

2003 Assessment	2005 Assessment					Total 2003
	VDF	MDF	OF	Scrub	NF	
VDF	17,470	23	2	0	21	17,516
MDF	11	78,349	67	16	807	79,250
OF	65	76	71,148	58	2,219	73,566
Scrub	0	0	56	440	588	1084
NF	0	9	2,778	52	80,828	83,667
Total 2005	17,546	78,457	74,051	566	84,463	255,083
Net Change	**30**	**-793**	**485**	**-518**	**796**	

It is obvious that the main cause of degradation or destruction of forests is nothing but biotic pressure. As the population increases, owing to our dependence on forests, biotic pressure on the forests also increases.

Logically, therefore, forests can be protected and conserved by reducing the biotic pressure only, which can normally be done either by controlling the population and/or by reducing the dependence on the forests.

As per the 2001 census, the following is the state wise population and their percentage decadal growth rate of the north east region during the decade from 1991 to 2001 (*Statistical Hand Book of Assam, 2006).*

State	Population in 2001 (in lakhs)	Per centage decadal growth rate
Arunachal Pradesh	10.98	27.00
Assam	266.56	18.92
Manipur	22.94	24.86
Meghalaya	23.19	30.65
Mizoram	8.89	28.82
Nagaland	19.90	64.53
Tripura	31.99	16.03
Total	384.45	22.56

Population growth is dependent on many factors, yet there is definite need to control the same, if biotic pressure on the forests need to be reduced. Under no circumstances, encroachment in the forests should be tolerated.

For reducing the biotic pressure on the forests by way of reducing our dependence on them, we will have to adopt integrated approaches, wherein use of substitutes of the materials collected from forests will have to be encouraged and also alternative livelihoods will have to be provided to the people dependent on forests for reducing their pressure on the forests.

The system of Joint Forest Management under the principle of Participatory Management has got the capability of bringing in the sense of ownership of the forests in the minds of the managing community and hence can protect the forests from the smugglers and encroachers, in addition to improving their conditions by raising successful plantations of suitable species. However, the sustainability of the system would depend only on the continuance of the physical and tangible benefits, which the community will be able to derive directly from the said forests. Once the physical benefits start reducing, there is likelihood that the people will also start losing interest on the protection of the forests and may try to extract things even at the cost of the forest itself. Hence the strategy and planning will have to be such that the community must be able to get their physical benefits from the forests on a sustainable basis.

As regards the use of substitutes of the materials collected from forests and also developing alternative livelihoods for reducing dependence on forests, the objective of this talk is not to specify the options in details, as the same are commonly available in many places. What is required is a concerted effort on the part of the authorities concerned and the awareness of the common people. What is felt is that a massive drive is required to make the people aware about the seriousness of the situation and their duties and responsibilities to avert the impending disaster, which can be done through education, media publicity, public meetings, organizing workshops, seminars etc. The students, public leaders and the women folk are felt to be the priority target groups in such campaign, who will be able to create a lot of impact on the society, once they themselves visualise and get convinced about the situation.

References

Forth Survey of India Report, 2005.

Statistical handbook of Assam, 2006.

Forest Survey of India Report, 2005.

12

Human Intervention: Effects on Forest and Wildlife in Meghalaya

Surajit Sen

It is the common understanding of *natural environment* that underlies *environmentalism* - a broad political, social, and philosophical movement that advocates various actions and policies in the interest of protecting what nature remains in the natural environment, or restoring or expanding the role of nature in this environment. While true wilderness is increasingly rare, *wild* nature (e.g., unmanaged forests, uncultivated grasslands, wildlife, wildflowers) can be found in many locations previously inhabited by humans.

Goals commonly expressed by environmental scientists include:

- Reduction and clean up of pollution, with future goals of zero pollution;
- Cleanly converting non-recyclable materials into energy through direct combustion or after conversion into secondary fuels;
- Reducing societal consumption of non-renewable fuels;
- Development of alternative, green, low-carbon or renewable energy sources;

- Conservation and sustainable use of scarce resources such as water, land, and air;
- Protection of representative or unique or pristine ecosystems;
- Preservation of threatened and endangered species extinction;
- The establishment of nature and biosphere reserves under various types of protection; and, most generally, the protection of biodiversity and ecosystems upon which all human and other life on earth depends.

Very large development projects - mega projects - pose special challenges and risks to the natural environment. Major dams and power plants are cases in point. The challenge to the environment from such projects is growing because more and bigger mega projects are being built, in developed and developing nations alike.

Forests, as is known, influences climate and moderate temperature and atmospheric humidity. It absorbs atmospheric carbon dioxide and replenishes oxygen; regulates stream flows by intercepting rainfall and gradually releases it into streams and rivers; regulates downstream flooding and drought conditions; provides shade and reduces evaporative loss from radiant energy exchange with the atmosphere; and provides shelter to all species of fauna besides being the host to numerous flora types. However, growing population, human greed and lack of knowledge about ecological balance have resulted in large scale destruction of flora and fauna.

It is in this context that the paper tries to evaluate the effect of human intervention upon the forest and wild life species of Meghalaya.

Meghalaya covers an area of 22,429 square kilometers. It has a total forest cover of 9,496 square kilometers. (FSI Report, 2003). It is home to an array of vegetation and wildlife like:

- Sal, Pine, Teak, Bamboo, Canes, etc
- Bio-genetic plants like Blue Vanda, Dendrebium, etc
- Fauna like the Royal Bengal Tiger, Elephant, Hornbill, Partridge, etc

Such a myriad collection of species is only possible due to the diverse topography, varied rainfall, differential climate and edaphic conditions. However, all these species are in constant threat for their very existence in the state of Meghalaya due to human intervention, the chief among the various causes could be:

- Agricultural activities in forest like *Jhuming*
- Over-exploitation of forest for firewood
- Expansion of animal farming land
- Industry and industrial purposes – Mining
- Other developmental; activities – Construction of Roads

To access the scenario on a broader basis, it is necessary to fragment the causes and species upon which the effect of human intervention is the most.

Effect of *Jhuming*

In *Jhuming*, there is a frequent shifting from one land to another affecting the ecology. Natural forests have declined due to constant cutting and burning of forests, and the repeated short-cycle *Jhuming* has created forest-canopy gaps which are evident form the barren hills. This has led to fragmentation of habitat and local disappearance of native species, and invasion by exotic weeds and other plants.

Effect on Flora

The famed *Khasiana Nepenthes* (Pitcher Plant) found only in Meghalaya is under threat due to random, rampant, unscientific practices of coal mining in the Jaintia Hills. It is known that *Orchids* are a special attraction in Meghalaya. About 300 species of orchids are found here. However, they are being destroyed heavily for commercial purposes. The point to be noted here is that orchids are regarded as keystone species in a forested ecosystem. Many organisms like insects, moths and even birds, depend on orchids for nectar and in return they help the plant in pollination. Depletion of orchids form an area can cause the depletion in population of these associated organisms. Mining also poses threat to orchids, especially in Garo Hills. With the shrinkage of green cover everywhere, rhododendrons are also facing the impact of a disturbed ecological system. Major threats to rhododendrons in Meghalaya are the expansion of agricultural activities in the forest, expansion of animal farming land and extensive collection of flowers.

Effect on Fauna

Garo Hills in Meghalaya has one of the largest and densest *Asian Elephant* populations in India. The principal threat to elephants in Garo Hills arises from large scale conversion of land use to such forms that are not compatible with long-term survival of elephants. With declining viability of *Jhum*, farmers are increasingly replacing the *jhum-fallow* complex with large scale plantations, which have obvious negative consequences for wildlife. There exists scanty and unreliable information on elephant numbers and absolutely no information on their ranging patterns in the landscape. The absence of such information makes it difficult to undertake land use planning with local communities since elephant movement cannot be factored into the land use plans. The framing of land use plans (and subsequent livelihood strengthening activities) in the absence of such information would potentially exacerbate conflict and continue fragmentation of elephant habitats. The threat to other animals is also no less potent. Environmentalists are out against coal mines alleging it to threaten the inmates of the *Balpakram National Park*, particularly *tigers*, *leopards*, *elephants*, *gibbons*, *deer*, *great Indian hornbill*, *pheasant* and *wild cats*. The *Siju Bird Sanctuary*, a host to many bird species, is under threat due to coal mining and road construction in Garo Hills.

Construction of Road

In recent times it has been noted that the amount of sediments eroded away from sites of road construction is ten times greater than that derived from agricultural land, about 200 times greater than that from grassland and 2000 times greater than that from forest land. Construction of roads disturbs the stability of the hillside, causes serious damage to the hydrologic system and removes the protective cover from the vulnerable slopes. Since 1971, there has been shrinkage of reserved forests of about 456.06 ha, mainly due to construction of roads. Road construction results in initiation of phases of persistent natural hazard, especially landslides that are alarm signals for mass extinction of many important plants including economically important *bamboo*, *cane* and *medicinal plants* along with other major *timber-yielding plants*. Many *ferns* are found damaged due to massive erosion of soil.

Industry versus Environment

In Meghalaya, two cement factories have recently come up around the limestone caves in *Lumshnong* and in the vicinity of the *Narpuh* reserved forests of Jaintia Hills. Both plants involve investments above Rs 100 crore. According to norms laid out by the ministry of environment and forests vide their environmental impact assessment (EIA) notification dated 1994, cement plants above Rs 100 crore investment should not be set up at a place which is –

(a) Less than 5 km (aerial distance) from reserved forests or wild life sanctuaries;

(b) Less than 5 km aerial distance from place of Tourist or Cultural or Historical importance; and

(c) Less than 5 km aerial distance from the international or state borders.

These factories have scrapped several plants and caused critical damage to wildlife habitats. The limestone mining above the *Siju* caves, a home to *Macrobrachum cavernicola*, a cavernicolous shrimp, is under serious threat. Three species of bats are also found in the caves: the *Long Winged Bat*, *Dobson's Long Tongued Fruit Bat* and the *Kashmir Cave Bat*, which are also under threat.

Unscientific Coal Mining

Unscientific Coal mining can result in number of adverse affects on the environment. In Meghalaya, mining of coal is mostly carried out in the Jaintia and Garo Hills. The method used for mining is highly traditional and unscientific. It is termed as *Rat Hole Mining*. The surface mining of coal completely eliminates existing vegetation, destroys the genetic soil profile, displaces wildlife and habitat, degrades air quality required for all forms of flora and fauna, alters current land uses, and to some extent permanently changes the general topography of the area mined.

Construction of Dams

In view of the proposed industrialization of the state and the entire northeastern region, the Meghalaya government and the North eastern Council has taken steps to develop thermal as well as hydro power generating stations in Meghalaya and the state is looking for investments in these sectors. One of the projects being taken up is the *Myntdu-Leshka* hydroelectric project stage – I (84 MW) in the Jaintia hills of Meghalaya. The project will totally require 181.16 ha. of land, of which 102 ha. is forestland. The endangered wildlife recorded in this area includes *tiger*, *jungle cat* and *binturong*, among other species, which will certainly be affected by this project.

Therefore, preserving biodiversity should be taken into consideration. Population growth, rapid expansion of cities and other settlements, land for new agricultural farms, mining, construction of village roads, dams, etc. have resulted in large-scale destruction of natural vegetation. However, proper management action plan by Government/semi-government/ local agencies is required at this juncture to protect the existing pristine environment of Meghalaya.

References

Assam Tribune, "Drive to Save Forests in Garo Hills", July 26, 2006.

Assam Tribune, "Exotic Flora in Meghalaya is Under Threat", November 9, 2006.

Conservation of Endangered Wild Life, Kalpana Palikhiwala.

Current Science, Vol. 91, No. 3, August 2006.

Current Science, Vol. 95, Nos. 7 & 10, October 2008.

Ecological Problems due to Shifting Cultivation, R. Ranjan and V. S. Upadhayay.

Meghalaya Biodiversity Conservation and Dams, B. Kharbuli, H. Kayang, D. Nengnong, D. R. M. Buam.

Ministry of Environment and Forests, Press Releases, June 26, 2009.

Newsletter, Soil and Water Conservation Meghalaya, Vol. 6, No. 6, July – December 2003.

Newsletter, Wildlife Institute of India, Vol. 13, No. 1, Spring 2005.

Paryavaran Abstracts, Vol. 19, Nos. 3 & 4, 2002.

Samrakshan Trust Meghalaya, December 2006.

The Financial Express, "Government Pressure: A Threat to N-E Forest", December 5, 2004.

The Pioneer, "Meghalaya to Clean up Coal Mining Act", August 1, 2006.

The Telegraph, "To Mine or not to Mine", November 20, 2003.

http://databank.nedfi.com

http://meghalaya.nic.in

http://ne.icrindia.org

http://www.indopia.in

http://www.intoday.in

http://www.megsoil.gov.in

13

Poverty-Environment Linkage in Relation to Sustainable Development: An Analysis in the Context of North East India (With Special Reference to Forest Resources)

M. B. Das
H. Choudhury
M. Dasgupta

Poverty is a global phenomenon. Three quarter of the people worldwide are living below the poverty line measured at an income of \$1/\$2 per day. Human poverty, today, is a multi-dimensional concept. It reflects inequitable access to the means of production, health care, education, employment and other services. According to Human Development Report (1997), poverty is the denial of opportunities 'to lead a long, healthy, creative life and to enjoy a decent standard of living, freedom, dignity, self-respect and the respect of others'.

Although the world has witnessed rapid development, many countries, especially the developing countries like India, continue to be plagued by the existence of poverty. Despite India's impressive growth rate in the last decade, the benefits have not percolated to the poor. This is mainly due to the failure of the distributional pattern of growth. The spillover effect of this is the co-existence of poverty, rapid population growth and environmental degradation. They seem to not only co-exist but also reinforce each other.

This has motivated us to analyze the interrelationship between poverty and environment. In this paper we study the linkage with special reference to the North East Region of India. Due to non-availability of relevant data, Sikkim has been excluded from the purview of the study. The rest of the paper is laid out in the following manner. In section 2, we discuss the poverty- environment nexus. A brief description of the North East Region (NER) is given in section 3. In section 4, the linkage between poverty and environment is studied with reference to the region. Estimates to prove the link is reported in section 5. Conclusion with suggestive policies to ensure sustainable development of the region is given in the last section.

Poverty-Environment Nexus

The Earth's environment is changing due to enormous increase in human activities at all levels- local, regional and global. Relentless pursuit of industrial growth has resulted in the pollution of air and water, depletion of natural resources and destruction of natural beauties. The focus of development has, therefore, shifted from growth to the concern for "quality of life" and sustainable development. All development activities are now aimed at "meeting the needs of the present without compromising the ability of the future generations to meet their own needs". This has become necessary to make development sustainable. Sustainable Development, thus, calls for equitable distribution and access to goods, services and natural Environment and Development (1987), in today's world, poverty is a major cause and effect of global environmental problems. Poor people are more vulnerable to environmental changes because of their low adaptive capacity. Many countries today, are caught in a downward spiral, that is, poor people are compelled to overuse natural resources for their existence and this jeopardizes the environment, which in its turn, leads to their further impoverishment making their lives more difficult (WECD1987). This inter-relationship leads us to believe that as poverty increases, environmental degradation starts increasing which results in a further decline in livelihood.

The relationship between poverty and environment cannot be viewed as an uni-dimensional cause-effect relationship. There exists two way causality between them. It is pertinent to note that the link between poverty and environment is influenced by the growth of population. High growth rate of population adversely affects environment via poverty. The poor with generally high fertility rates, low education and health care facilities exert more pressure on the natural resources to meet their requirements. This results in environmental degradation, which in turn, enhances the hardship of the poor. For instance, fuel wood crisis, shortage of drinking water, soil erosion and flooding that follows deforestation, loss of fisheries, soil degradation etc., affects the poor more than the rich. Environmental problems and poverty, thus, have been conceived as being reinforcing and reciprocally inter-related (World Bank, 2001).

The above explained relationship is depicted in the following flow chart.

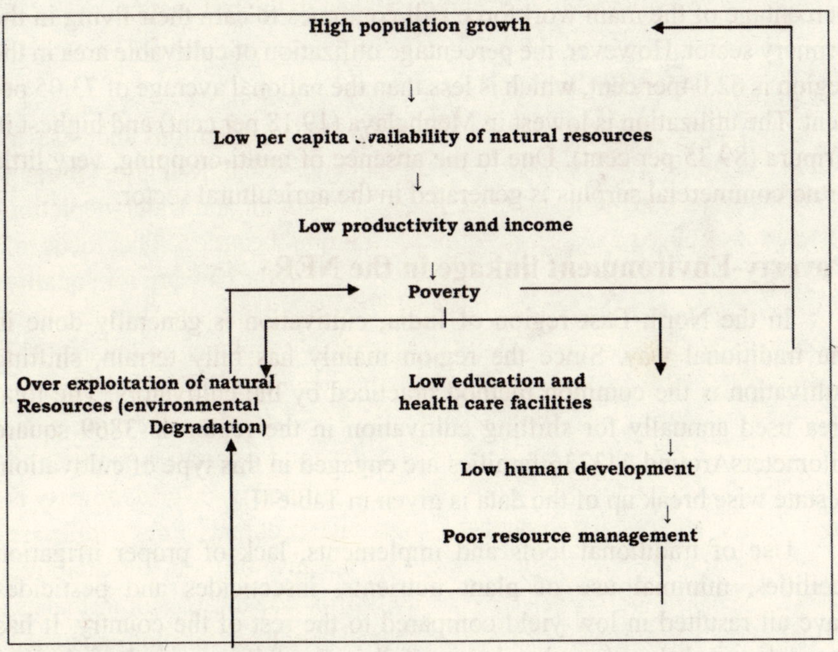

The North East Region of India

The North East Region (NER) of India accounts for nearly 8 per cent of the country's geographical area. The region, comprising of the states of Arunachal Pradesh, Assam, Manipur, Meghalaya, Mizoram, Nagaland, Sikkim and Tripura, possesses many unique features making it one of the most beautiful geographical regions of the country. The region is not only diverse in terms of topography and natural resources but also with regards to heritage and culture. The numerous varieties of flora and fauna of the region make it one of the sixteen bio-diverse regions of the world.

According to the 2001 Census, the region accounts for 3.75 per cent of the total population of the country. The growth rate of population of the region, however, is higher than that of the country. The decadal growth rates reveal that population of the North East Region was 27.46 per cent in 1981-1991 and 22.02 per cent in 1991-2001 as against the all India growth rates of 23.86 per cent and 21.34 per cent during 1981-1991 and 1991-2001 respectively.

The North East economy is predominantly agrarian by nature. A glimpse at Table I reveals that although there has been a gradual increase in the percentage of main workers in the tertiary sector, a very significant percentage of the main workforce still continues to earn their living in the primary sector. However, the percentage utilization of cultivable area in the region is 62.04per cent, which is less than the national average of 73.05 per cent. The utilization is lowest in Meghalaya (19.18 per cent) and highest in Tripura (89.35 per cent). Due to the absence of multi-cropping, very little or no commercial surplus is generated in the agricultural sector.

Poverty-Environment linkage in the NER

In the North East region of India, cultivation is generally done in the traditional way. Since the region mainly has hilly terrain, shifting cultivation is the common method practiced by the cultivators. The total area used annually for shifting cultivation in the region is 3869 square kilometersAround 443336 families are engaged in this type of cultivation. A state wise break up of the data is given in Table II.

Use of traditional tools and implements, lack of proper irrigation facilities, minimal use of plant nutrients, insecticides and pesticides have all resulted in low yield compared to the rest of the country. It has been observed that there has been a fall in the fallow period of shifting cultivation from 15-20 years to 3-4 years. This may be attributed to the excessive pressure of population on land and consequent decline in the per capita availability of land. Table III shows that the per capita availability of land has declined from 1.16 square kilometers in 1971 to 0.58 square kilometers in 2001. The reduction in the cycle period of shifting cultivation has reduced the fertility of the soil resulting in low yield and low income thereby adding to the miseries of the people especially the poor.

The practice of shifting cultivation also adversely affects the environment. It leads to large scale deforestation and loss of vegetation and bio-mass. Since shifting cultivation is practiced on hill slopes, downstream siltation is apparent in many places. These areas receive moderate to heavy rainfall. The splash forces generated by the rain also cause erosion of the top soil.

Lack of industrialization, low infrastructure and insufficient employment opportunities have forced the poor to depend mainly on agriculture and allied activities for livelihood. The poor are compelled to use the marginal or infertile land for fulfilling their sustenance requirement. Over exploitation of such marginal land result in further decline in productivity and income causing more poverty. Thus a cyclical relationship between poverty and environmental degradation develops.

Again, the poor have to depend on bio-fuel to meet their energy requirement since they do not have enough resource to use eco-friendly commercial fuel. For instance, the poor people of the region generally use wood for cooking. Increasing demand for firewood is resulting in large scale deforestation. Since the forests of the hill states generally belong to the local communities, they are easily available. Large scale deforestation also takes place for extracting timber for commercial purpose. Table IV shows the net change in forest cover in square kilometersand the extent of loss due to shifting cultivation and other reasons.

Timber and various forest products also form a principal source of livelihood for the poor. Various herbs and plants, besides meeting the food requirement are also a source of revenue for the poor. Depletion of forest resource not only lead to energy, food and revenue crisis for the poor, but also result in poor rainfall causing shortage of water both for the purpose of cultivation and drinking. These areas become vulnerable to water logging, endemic floods and landslides causing distress to the people and more so to the poor. With environmental degradation, the prospects of further livelihood declines generating more poverty which in turn exerts more pressure on the environment, thereby completing the vicious cycle.

It may also be mentioned here that the use of wood as a fuel is also a source of environmental pollution. The carbon-dioxide formed due to the burning of wood pollutes the air and adversely affects the health of the poor living in slums and poorly ventilated houses. Thus, it may be inferred that environmental pollution by adversely affecting the health of the poor results in a decline in efficiency which in turn leads to low productivity and low income.

Open and unscientific mining of coal, limestone etc, also result in environmental degradation. For the purpose of mining, the mineral rich areas are cleared of valuable trees and vegetation resulting in large scale deforestation. The soil removed for mining is dumped in nearby areas creating artificial hillocks of rocks and debris in place of natural green hillocks. Because of mining these areas become susceptible to frequent landslides.

Relationship among poverty, per capita income and changes in forest cover

Though the poverty alleviation initiatives taken by the State Governments have resulted in a decline in the incidence of poverty over the years, 1999-2000 estimates reveal that all the states of the region , except Mizoram had poverty ratios above the national average 26.10per cent. Table V shows the variance in poverty ratios among the states of the region with Assam recording the highest ratio.

Table VI shows that there was a net loss in forest cover in all the states, except Mizoram and Tripura, during 1995-97. Assam recorded the highest amount of deforestation. Correlation between incidence of poverty and change in forest cover has been found to be -0.751 and is significant. Significant negative correlation (-0.754 and -0.790) also exists between change in forest cover and percentage changes in poverty ratios.

These estimates corroborate the fact that an inverse relation exists between poverty and change in forest cover. More poverty implies larger decline in forest cover which means an increase in environmental degradation. The results, thus, support the inverted U relationship that in the early stages of development , environmental degradation does take place.

The correlation between per capita income and change in forest cover has been found to be positive (0.468) but not significant. This implies that a change in per capita income has no direct relationship with the area and volume of forests.

Discussion

The present study reveals that a vicious cycle of poverty and environmental degradation does operate in the region. This is impeding the level of development. Large scale depletion of natural resources , especially the exhaustible resources , for meeting the present requirements of an ever increasing population is impairing the future prospects of development of the region. The question is: how to break the cycle? The answer lies in the sustainable development approach. Sustainability refers to both economic and ecological attributes. Mere increase in per capita SDP cannot ensure sustainable development. To be sustainable, development should bring about an improvement in overall wellbeing of the people while 'ensuring that stream of future consumption rates to be at least the same as of now'.

Efforts to reduce poverty through poverty alleviation programmes alone will not suffice. Redistributive policies with strong institutional support needs to framed. What is needed at the moment is a sustainable livelihood approach through improved access to high quality technical and vocational education, better nutrition , health care, more facilitating infrastructure, access to better financial assistance, cohesive and supportive social environment etc. Environmental sustainability can be achieved through conservation, human development, proper knowledge about environmental care, community based approach towards environment protection, a change in taxation and pricing policies, market based approach for preservation of environment etc.

The State Governments should refrain from using forests as a revenue earning resource. Policies aimed at restoration of ecological balance through afforestation should be framed and implemented. Preservation of the natural forests with numerous varieties of flora and fauna will keep the biodiversity and genetic resource of the region intact. The National Forest and Wildlife Protection Policy of the government of India should be strictly adhered to. The State Governments should play a catalytic role in the management of village commons and forest lands.

Besides controlling the rapid growth of population, the State Governments should also undertake land reform measures to reduce the imbalance in the distribution of land especially in the rural areas. Alternative employment opportunities should be provided to people so that over-exploitation of natural resources is reduced. Proper implementation of policies and programmes will ensure growth, reduce poverty and protect and preserve the environment leading to the sustainable development of the region.

Table: 1 Changes in the Sectoral Distribution of Main Workers in NER

States	Percentage of Main workers in Primary Sector		Percentage of Main workers in Secondary Sector		Percentage of Main workers in Tertiary Sector	
	1971	1991	1971	1991	1971	1991
Arunachal Pradesh	80.44	67.44	0.45	8.66	19.19	23.96
Assam	77.04	73.99	0.20	5.56	17.86	20.45
Manipur	71.30	70.70	12.24	9.66	16.46	20.34
Meghalaya	81.84	74.81	3.30	3.74	14.86	21.46
Mizoram	84.17	67.99	1.76	5.07	14.07	28.94
Nagaland	79.46	75.26	2.38	3.48	18.16	21.26
Tripura	76.58	64.08	4.25	6.41	19.17	29.51

Table 2: State-wise break-up of Shifting cultivation practiced in NER

State	Annual area under Shifting Cultivation (sq. kms.)	Minimum area under shifting cultivation at one time or other (sq. kms.)	Number of families practicing shifting cultivation
Arunachal Pradesh	700	2,100	54,000
Assam	696	1,392	58,000
Manipur	900	3,600	70,000
Meghalaya	530	2,650	52,290
Mizoram	630	1,890	50,000
Nagaland	190	1,913	1,16,046
Tripura	223	1,115	43,000
NER	3869 (1.5per cent)	14,660 (5.7per cent)	4,43,336

N.B. 2001 data is unavailable.

Source: Basic Statistics of NER, 2006

Table: 3 Per-capita availability of land in NER

State	1971	1981	1991	2001
Arunachal Pradesh	12.07	8.86	6.36	5.04
Assam	0.54	0.39	0.35	0.29
Manipur	2.206	1.56	1.20	0.93
Meghalaya	2.22	1.68	1.26	0.97
Mizoram	6.52	4.26	3.03	2.235
Nagaland	3.20	1.79	1.28	0.79
Tripura	0.67	0.51	0.38	0.33
NER	1.16	0.84	0.71	0.58

Source: Basic Statistics of NER, 2006 and NEDFi Report

Table: 4 Change in forest cover in NER (sq. kms.)

States	Net change in forest cover		Loss due to shifting cultivation		Loss due to other reasons	
	1995-97	2001-03	1995-97	2001-03	1995-97	2001-03
Arunachal Pradesh	-19	-1026	+75	+925	N.A.	+256
Assam	-237	+112	+257	+609	+159	+2086
Manipur	-140	+293	+603	+855	N.A.	+108
Meghalaya	-57	+1255	+75	+684	+2	+71
Mizoram	+199	+936	+292	+687	N.A.	+900
Nagaland	-70	+264	+573	+1332	N.A.	+57
Tripura	+8	+69	N.A.	+384	+3	+1

Source- Forest Survey of India, Dehradun

Table: 5 Poverty ratios in North-East India

State	Poverty Ratio in 1999-2000 (per cent)	Per Capita Income 1999-2000 (at 1993-94 Price)	Changes in Forest Cover in 1995-97 (sq. kms.)	Percentage Changes in Poverty Ratios	
				1993-94 to 1999-2000	1983 to 1999-2000
Arunachal Pradesh	33.47	9170	-19	-14.94	-17.11
Assam	36.09	5978	-237	-11.67	-10.82
Manipur	28.58	7231	-140	-15.51	-22.91
Meghalaya	33.87	7826	-57	-10.68	-12.73
Mizoram	19.47	8109	199	-24.12	-45.92
Nagaland	32.67	9118	-70	-13.84	-16.76
Tripura	34.44	6813	8	-11.71	-13.96
	R_{34}=0.468	R_{24}=-0.751	R_{45}=-0.754	R_{46}=-0.790	

Source- Economic Survey, India 2001-02

Table: 6
Poverty, Per-capita Income and Changes in Forest Cover in NER

States	Poverty Incidence in Percentage			Percentage changes in Poverty Ratios		
	1983	1993-94	1999-2000	1983 to 1993-94	1993-94 to 1999-2000	1983 to 1999-2000
Arunachal Pradesh	40.38	39.35	33.47	-2.55	-14.94	-17.11
Assam	40.47	40.86	36.09	0.96	-11.67	-10.82
Manipur	37.02	33.78	28.54	-8.75	-15.51	-22.91
Meghalaya	38.81	37.92	33.87	-2.29	-10.68	-12.73
Mizoram	36.00	25.66	19.47	-28.72	-24.12	-45.92
Nagaland	39.25	37.92	32.67	-3.39	-13.84	-16.76
Tripura	40.03	39.01	34.44	-2.55	-11.71	-13.96

Source- Directorate of Economics and Statistics of respective State Governments; Forest Survey of India, Dehradun.

References

Bandyopadhyay, Jayanta (1998): "Approach to Sustainable Human Development in the Indian Himalaya, Concept paper, IIM, Calcutta, India.

Barthakur, D. (1997): "Jhuming and consequences", Yojana, Annual Number, XXI,1: 122-123.

Baruni, Tariq (1998): "The Case of the Environmental Kuznets Curve", Background Paper for "Human Development Report 1998", New York, UNDP.

Beckman, M. and Pun, T. (1985): "Spatial economics : Density, Potential and Flood, Studies in Regional Sciences and Urban Economics, Amsterdam, North Holland.

Bhattacharya, R.N. (ed) (2001): Environmental Economics: An Indian Perspective". Oxford University Press, New Delhi.

Chopra, Kanchan and Gulati,S.C. (1998): "Environmental degradation , Property Rights and Population Movements: Hypotheses and evidence from Rajasthan, India', Environment and development economics, no. 3, pp 35-57, Cambridge University Press.

Dasgupta, P.(2003): "Population, Poverty and the Natural Environment", in K.G.Maler and J.R.Vincent(eds): Handbook of Environmental Economics , vol. I, Edward Elgar, Cheltanham, pp 191-247.

Hanley, Nick, J. Shogren and Ben White (1999): Environmental Economics, New Delhi, Macmillan.

Jodha, N.S.(1986): "Common property resources and rural poor in dry regions of India", Economic and Political Weekly, Vol.21, no.27, pp 169-181.

Jodha, N.S (2000): "Waste lands management in India: Myths, Motives and Mechanisms", Economic and Political Weekly, Feb 5, pp 466-473.

Kadekodi,G.K. (2001); "Environmental Degradation and Resilience to Human Development: A Case Study of Karnataka", In A.P. Bali(ed), Refashioning the new Economic order: Karnataka in Transition, ICSSR and Rawat Publications, New Delhi.

Kuznets, S. (1955): "Economic Growth and Income Inequality", American Economic Review, 49.

Markandya, A.(1998): "Poverty, Income Distribution and Policy Making', Environmental and Resource Economics, vol.II(3-4): pp 459-472.

Ministry of Agriculture, (1999): Agricultural Statistics at a glance-1998, directorate of Agriculture and Co-operation, Ministry of agriculture, New Delhi.

Ministry of Environment and Forest, (19970): Forestry Statistics of India, 1988-94, Ministry of Environment and Forests, Directorate of Statistics, Indian Council of Forestry Research And Education, Dehradun.

Oates, W.E.(ed) 91992): "The Economics of Environment", Brookfield, Edward Elgar.

Pearce, D. and R.K.Turner (1990): Economics of Natural Resources and the Environment, New York: Harvester and Wheatcheef.

Solow, R.M. (1974): "The Economics of Resources or the Resources of Economics", American Economic Review, 49.

World Commission On Environment and Development (1987): "Our Common future", New York, Oxford University Press.

Wasteland Atlas of India, Ministry of Rural Development, Government of India and National Remote Sensing Agency, 2000.

World Bank (1979): World Development Report, 1997, Washington D C.

World Bank Report (2000): Bangladesh Study of Land Issues, SASRD, World Bank Dhaka Office, 2000.